T0243097

Monograph Series Volume 4

Cases in Mathematics Teacher Education: Tools for Developing Knowledge Needed for Teaching

Edited by

Margaret S. Smith
University of Pittsburgh

Susan N. Friel
University of North Carolina – Chapel Hill

Monograph Series Editor

Denisse R. Thompson
University of South Florida

Association of
Mathematics Teacher Educators

Published by the Association of Mathematics Teacher Educators, San Diego State University, c/o Center for Research in Mathematics and Science Education, 6475 Alvarado Road, Suite 206, San Diego CA 92129.

www.amte.net

Library of Congress Cataloging-in-Publication Data

Cases in mathematics teacher education : tools for developing knowledge needed for teaching / edited by Margaret S. Smith, Susan N. Friel.
p. cm. -- (Monograph series ; v. 4)
Includes bibliographical references.
ISBN 978-1-62396-947-9
1. Mathematics teachers--Training of. 2. Mathematics--Study and teaching.
3. Case method.
I. Smith, Margaret Schwan. II. Friel, Susan N.
QA10.5.C376 2008
510.71--dc22

Library of Congress Catalog Control Number: 2008003558
International Standard Book Number: 1-932793-05-4

Contents

Bay-Williams, J. M.
AMTE Monograph 4
Cases in Mathematics Teacher Education: Tools for Developing Knowledge Needed for Teaching
©2008, pp. v-vi

Foreword

The goal of AMTE Monograph 4, *Cases in Mathematics Teacher Education: Tools for Developing Knowledge Needed for Teaching,* is to provide detailed accounts of case use that will inform the mathematics teacher education community on the range of ways in which cases can be used to foster teacher learning and the capacity to reflect on and learn from teaching. The chapters in this monograph describe the use of cases with preservice and practicing teachers at all levels K-12, in content and methods courses as well as professional development settings, and focus on developing various aspects of teachers' knowledge base (i.e., content, pedagogy, and students as learners). Hence, Monograph 4 should prove to be a superb resource for mathematics teacher educators.

This monograph has a unique history that stands as a model of the workings of AMTE, so I will share this background briefly and how it exemplifies the work of the association. The monograph did not get its beginnings by some members thinking what the next Monograph topic might be; this monograph got its start as an AMTE Board conversation about resources that could be useful to AMTE members. The notion emerged that offering bibliographies on specific mathematics teacher education topics would be a great help. Margaret Smith and Susan Friel, Members-at-Large on the Board of Directors at the time, agreed to prepare the first one, which would be focused on cases.

Not long after this work began, a Teaching Resources Task Force was formed to strategize what would be the most useful and effective product for AMTE members in terms of teaching/professional development resources. Naturally, Peg and Susan became the Co-Chairs of the Task Force. Collectively they decided that both the needs of AMTE members and the value of cases for teacher learning would be better served with a full collection of articles of cases used in real professional education settings. A call was developed and the rest of the story is found among the pages of this monograph. What started as a small, short-term project took on a life of its own - resulting in the production of this resource. Hence the monograph serves as evidence of the initiative, vision, hard work, and continued effort of members of AMTE.

In addition to projects initiated by the AMTE Board, the organization is extraordinary in that its members, committees, or Task Forces bring forth ideas, persist in their development, and produce great work that better supports the members of AMTE. To the contributing authors, thank you! Sharing your work using cases will be greatly appreciated and used by many. To the reviewers and editors, thank you! Your efforts have enhanced the quality of the manuscripts and resulted in an overall excellent resource.

To those who participated in the Teaching Resources Task Force, thank you for your hand in the creation of Monograph 4:

Margaret Smith, University of Pittsburgh, PA, Co-Chair
Susan Friel, University of North Carolina-Chapel Hill, NC, Co-Chair
M. Lynn Breyfogle, Bucknell University, PA
Susan Hillman, Saginaw Valley State College, MI
Amy Roth McDuffie, Washington State University – Tri Cities, WA
Babette Moeller, Bank Street College, NY
Kathy Morris, Sonoma State University, CA

Collectively, the authors, editors, editorial panel, and Task Force members have prepared a resource that provides a breadth and depth of knowledge on using cases – certainly a resource that is and will continue to be of great value to mathematics teacher educators.

Jennifer M. Bay-Williams
AMTE President 2007-2009

Smith, M. S. and Friel, S. N.
AMTE Monograph 4
Cases in Mathematics Teacher Education: Tools for Developing Knowledge Needed for Teaching
©2008, pp. 1-8

1

Cases in Mathematics Teacher Education[1]

Margaret S. Smith
University of Pittsburgh

Susan N. Friel
University of North Carolina – Chapel Hill

Since Lee Shulman first proposed "case knowledge" as a component of "teacher knowledge" in 1986, cases have gained momentum as important tools in the professional education of teachers. As an antidote to what many see as an overly theoretical approach to teacher education, many teacher educators have been drawn to cases because they capture the complexity and authenticity of instructional practice. Unlike theories, propositions, principles, or other abstractions, the particularities of cases vividly convey the profusion of events that constitute the moment-by-moment lived experience of classrooms. Merseth (2003, p.xvii) argues "good cases bring a 'chunk of reality' into the teacher education classroom to be examined, explored, and utilized as a window on practice."

Although there are a variety of interpretations of what constitutes a case, Shulman (1986) argued that to call something a case is to make a theoretical claim – that is, any story that is called a case must be a case *of* something. Hence, not every video or narrative that portrays an aspect of classroom life qualifies as a case. A case must make salient some idea, principle, or theory that is central to mathematics teaching and learning more generally – that is, the particulars portrayed in the case must be instances of larger, more generalized ideas. In addition, a case should be specifically designed to stimulate engagement and discussion.

In Chapters 2 through 11 of this monograph, mathematics teacher educators provide rich illustrations of the ways in which they have used specific cases to help teachers develop their knowledge base for teaching (i.e., knowledge of content, pedagogy, and students as learners) and the capacity to reflect on and learn from teaching. In this chapter we provide a general overview of cases and their use in mathematics teacher education and highlight the contributions of the individual chapters.

The *What* and *Why* of Cases

Materials that can effectively support teachers as they develop their pedagogical practice must do more than transmit general mathematical and pedagogical propositions that apply broadly across a range of situations. They must assist teachers to develop a capacity for knowing when and how to apply such knowledge, a capacity that depends on the ability to connect the specifics of real-time, deeply contextualized teaching moments with a broader set of ideas about mathematics, about teaching, and about learning. To develop this capacity, teachers must learn to recognize events in their own classrooms as instances of larger patterns and principles. Then they can formulate ways of acting and interacting that are thoughtful, principled, and effective (Shulman, 1996). As Shulman has noted:

> Teachers learn quickly that the heart of teaching is developing the capacity to respond to the unpredictable. Teaching begins in design, but unfolds through chance. And cases – as the narrative manifestations of chance – offer teachers the opportunities to contemplate the variety of ways in which the unpredictable happens. Case-based teacher education offers safe contexts within which teachers can explore their alternatives and judge their consequences (1996, p. 214).

Case use in teacher education has drawn heavily from the experience of case use in the professional fields of law, medicine, and business (Merseth, 1996). Although these fields are quite diverse, there appears to be a general belief that cases can address effectively a common tension in the design of experiences for professional education. A professional education curriculum seeks both to deliver a theoretically based knowledge base and to teach reasoning skills and strategies for analyzing and acting professionally in novel settings. Such a curriculum is grounded in the obligation of professional education to prepare practitioners for a practice that is simultaneously routine and uncertain (Sykes & Bird, 1992). Both of these aspects of professional education seem well suited to cases.

Although it is unlikely that cases alone are sufficient as a source of professional learning (Patel & Kaufman, 2001), they can be a critical component of a curriculum for teacher education, providing a focus for sustained teacher inquiry and investigation (Ball & Cohen, 1999) and an opportunity to make connections to experiences (vicarious or lived) and to theoretical classifications and general principles (Shulman, 1996). According to Smith (2001), cases also create opportunities for teachers to begin to develop new visions of mathematics teaching and learning and provide a common experience for teachers to discuss, analyze, and reference.

Cases Materials

Cases can be divided into two broad categories, *exemplars* and *problem situations* (Carter, 1999). Exemplars can be used to exemplify a practice or operationalize a theory. They provide vivid images of teachers in real classrooms that ground abstract ideas related to content and pedagogy. For example, the cases discussed by Silver and his colleagues (Chapter 8) exemplify key features of instruction associated with the implementation of cognitively challenging mathematical tasks (Smith, Silver, & Stein, 2005a, 2005b; Stein, Smith, Henningsen, & Silver, 2000) and invite the reader to analyze the pedagogical moves made by the teacher and the ways in which these moves support or inhibit a student's learning of mathematics. These cases are intended to assist teachers to develop an understanding of mathematical tasks and how they evolve during a lesson and enhance their ability to reflect critically on their own practice guided by a framework based on these ideas (Stein et al., 2000).

Problem situations, on the other hand, can be used to examine the complexities of teaching and the problematic aspects of performance. They often provide dilemmas (either mathematical or pedagogical) to be analyzed and resolved. For example, the case discussed by Morris (Chapter 2) poses a situation that occurred during instruction that the teacher had not anticipated and invites the reader to explore different courses of action and to consider the trade-offs in selecting one over another (Barnett-Clarke & Ramirez, 2003). By learning how to analyze "messy and complex situations," teachers can learn to make well-informed decisions in their own classrooms (Barnett & Ramirez, 1996, p.11).

Regardless of the type of cases used, cases are intended to help teachers develop the knowledge and skills needed to respond to the complexities and demands of real-time teaching and begin to think like teachers. According to Richardson (1996, ix), one challenge of teacher education is to help teachers "begin to develop practical knowledge that will allow them to survive the reality of the classroom." Cases appear to be one way of facilitating teachers' development of this practical knowledge.

The first book of mathematics cases was published in 1994 (see Barnett, Goldenstein, & Jackson, 1994) and launched a new era in the education of teachers of mathematics. Since the publication of this volume more than a decade ago, many additional mathematics casebooks have been published (e.g., Merseth, 2003; Schifter, Bastable, & Russell, 2007; Seago, Mumme, & Branca, 2004; Stein, Smith, Henningsen & Silver, 2000). (See the Appendix for a more complete listing.) These casebooks vary greatly in terms of content focus (e.g., specific mathematical ideas, students' thinking about particular pieces of mathematics, the pedagogy used to support student learning of mathematics), grade level (e.g., elementary, middle, high school), type (i.e., narrative or video), and authorship (i.e., written by teachers describing their own practice vs. written by a third party describing some aspect of classroom instruction). Despite these differences, cases in mathematics education share a common feature of providing realistic contexts for helping teachers "develop skills of analysis and problem-solving, gain broad repertoires of pedagogical technique, capitalize on the power of reflection, and experience a positive learning community" (Merseth, 1999, pp. xi-xii).

Learning from Cases

Learning from cases is not self-enacting. Reading a case does not ensure that the reader will automatically engage with all the embedded ideas or will spontaneously make connections to his or her own practice. Most of the case materials currently available (see Appendix) provide suggestions to support a facilitator's use of the cases. In fact, several casebooks have companion facilitation guides that provide explicit and extensive suggestions regarding how to use the cases.

Although there is no standard protocol for how cases are to be used, many case authors suggest having teachers begin their work on a case by first solving the task on which the case is based. Steele (Chapter 6) makes this point, arguing that working on tasks and cases *together* enhances the learning of mathematical, pedagogical, and pedagogical content knowledge.

Group discussion, deliberation and debate, however, are key to learning from cases (Shulman, 1996). The success of a case discussion depends in large measure on the skill of the facilitator in managing active learners with multiple (and often conflicting) viewpoints and in highlighting the question, "What is this a case of?" This question is of critical importance in stimulating learners "to move up and down, back and forth, between the memorable particularities of cases and the powerful generalizations and simplifications of principles and theories" (Shulman, 1996). Like an experienced teacher, a facilitator must decide "when to let students struggle to make sense of an idea or problem…, when to ask leading questions, or when to tell students something" (NCTM, 1991, p. 38). The choices made by the facilitator have an influence on the direction of the discussion, on the depth and range of issues that are brought to the fore, and on the opportunities participants have to gain new insights, question current practices, and continue to learn and develop as professionals. Goldsmith and Seago (Chapter 11) highlight the role of the facilitator in keeping the discussion focused on the goals of the professional development session in which the case is being used and characterize the moves made by the facilitator that help to focus teachers' attention on key aspects of the case. The identification of these discussion moves is of critical importance in helping novice teacher educators begin to embrace the case approach.

Although the decisions made by the facilitator during the discussion are critical, Silver and his colleagues (Chapter 8) and Kazemi and her colleagues (Chapter 3) highlight the importance of the decisions made by the facilitator in selecting and sequencing cases that will engage teachers in discussing issues which the facilitators have identified as important aspects of teachers' learning. Hence, deciding what case to use for what purpose is critical. According to Sykes and Bird (1992), the selection and sequencing of cases with other elements of teacher education is a complex curricular issue. Ball and Cohen (1999) caution us to design professional education experiences so as to avoid "simply reproducing the kind of fragmented, unfocused, and superficial work that already characterizes professional development" (p. 29).

Research on Learning from Cases

Although there is considerable enthusiasm for using cases in teacher education, and many claims regarding the efficacy of this approach (e.g., Merseth, 1991; Sykes & Bird, 1992), establishing an empirical basis for these claims has been a slow process. In 1999 Merseth noted "the conversations about case-based instruction over the last two decades has been full of heat, but with very little light" (p. xiv). One coherent attempt to define an empirical basis for the use of cases in teacher education is the book entitled, *Who Learns What From Cases and How?: The Research Base for Teaching and Learning with Cases,* edited by Mary Ludenberg, Barbara Levin, and Helen Harrington (1999). The chapters in this book report the findings of a series of studies, mostly descriptive or naturalistic, conducted by the authors in an effort to determine what students enrolled in their teacher education courses learned. Although the book edited by Ludenberg and her colleagues and many other research studies that are identified by the authors of the chapters in this volume provide support for case methods, additional research is needed to explore issues of teacher learning (e.g., what do teachers learn from different types of cases and how they learn it) and how what teachers learn impacts their teaching performance. Chapter 5, written by Henningsen, provides one example of the way in which a case discussion can impact a teacher's practice.

The Case Chapters

The chapters presented in this monograph are a first step in examining the use of cases in mathematics teacher education and highlight the diversity of cases themselves and the contexts in which cases can be used. For example, Chapter 7 (Hillen and Hughes) provides a detailed accounting of the use of lengthy narrative cases (Smith et al., 2005b) in a graduate-level methods course, with preservice and inservice teachers spanning nearly every grade level K-12. By contrast Chapter 4 (Shifter and Bastable) describes the use of a short narrative case (Shifter et al., 2007) with inservice elementary teachers. Finally, Chapters 9 (Romagnano, Evans & Gilmore), Chapter 10 (Van Zoest & Stockero), and Chapter 11 (Goldsmith & Seago) highlight the use of video cases. These chapters illustrate that "case strategies" applicable to written cases generalize to video cases. Table 1 provides an overview of the chapters, including the setting in which the case was used, the grade level(s) of the participating teachers, and the source of the case(s) discussed.

Many of the chapter authors have provided examples of how a case can be used for a purpose that may be somewhat different from what the authors of the case materials might have intended. Generally, when casebooks are written, the authors think of them being used "in their *entirety.*" So, for many of the case resources found in the Appendix, it is not unusual for the case authors to provide guidance in how to use the materials as a stand-alone professional development program. Indeed, we all know instances of such use and how valuable this can be (Chapters 3, 10, and 11 in fact describe situations where the materials were used in their entirety). *Decomposing* case resources and making choices to limit use to just one or two cases naturally requires artful orchestration. Many of the chapters in this volume provide thoughtful reports of ways that single or a few cases have been extracted from more comprehensive resources and how these extracted cases have been carefully sequenced with other materials to help teachers develop their capacity to analyze and reflect on practice. Such discussions of case adaptation are certainly a "first" for the case literature. As readers of this monograph reflect on the uses of cases described here, we hope they will begin to consider additional ways in which cases might be used, particularly when it is not feasible to use an entire case resource. We encourage readers to continue the discussion and sharing that have begun with this set of manuscripts.

Table 1. Overview of each chapter in the monograph

Chapter	Chapter Author(s)	Setting	Grade Level of Teachers	Source of Cases Discussed
2	Morris	Methods Course (Preservice)	Elementary	Barnett-Clarke & Ramirez (2003)
3	Kazemi, Lenges, & Stimpson	Professional Development (Inservice)	Elementary	Schifter, Bastable, & Russell (1999)
4	Schifter & Bastable	Professional Development (Inservice)	Elementary	Schifter, Bastable, & Russell (2008)
5	Henningsen	Methods Course (Preservice)	Elementary	Smith et al. (2005b)
6	Steele	Graduate Methods Course (preservice & inservice)	Elementary Middle High	Smith et al. (2005c)
7	Hillen & Hughes	Graduate Methods Course (preservice & inservice)	Elementary Middle High	Smith et al. (2005b)
8	Silver, Clark, Gosen, & Mills	Professional Development (Inservice)	Middle	Smith et al. (2005a, 2005b); Stein et al. (2000)
9	Romagnano, Evans, & Gilmore	Content Course (Preservice)	Middle High	Seago, Mumme, & Branca (2004)
10	Van Zoest & Stockero	Methods Course (Preservice)	Middle	Seago, Mumme, & Branca (2004)
11	Goldsmith & Seago	Professional Development (Inservice)	Middle	Seago, Mumme, & Branca (2004)

References

Ball, D. L., & Cohen, D. K. (1999). Developing practice, developing practitioners: Towards a practice-based theory of professional education. In L. Darling-Hammond & G. Sykes (Eds.), *Teaching as the learning profession: Handbook of policy and practice* (pp. 3-32). San Francisco: Jossey-Bass.

Barnett, C., Goldenstein, D., & Jackson, B. (Eds.). (1994). *Fractions, decimals, ratios, and percents: Hard to teach and hard to learn?* Portsmouth, NH: Heinemann. (Two books: case book and facilitator's guide)

Barnett, C., & Ramirez, A. (1996). Fostering critical analysis and reflection through mathematics case discussions. In J. Colbert, K. Trimble, & P. Desberg (Eds.), *The case for education: Contemporary approaches for using case methods* (pp. 1-13). Boston: Allyn & Bacon.

Barnett-Clarke, C., & Ramirez, A. (Eds.). (2003). *Number sense and operations in the primary grades: Hard to teach and hard to learn?* Portsmouth, NH: Heinemann.

Carter, K. (1999). What is a case? What is not a case? In M. A. Lundeberg, B. B. Levin, & H. L. Harrington (Eds.), *Who learns what from cases and how?: The research base for teaching and learning with cases.* Mahwah NJ: Lawrence Erlbaum.

Lundeberg, M. A., Levin, B. B., & Harrington, H. L. (Eds.). (1999). *Who learns what from cases and how? The research base for teaching and learning with cases.* Mahwah, NJ: Lawrence Erlbaum Associates.

Merseth, K. K. (1991). *The case for cases in teacher education.* Washington, DC: American Association of Colleges of Teacher Education.

Merseth, K. K. (1996). Cases and case methods in teacher education. In J. Sikula, T. J. Buttery, and E. Guyton (Eds.), *Handbook of research on teacher education* (pp. 722-744). New York: Macmillan.

Merseth, K. K. (1999). A rationale for case-based pedagogy in teacher education. In M. A. Lundeberg, B. B. Levin, & H. L. Harrington (Eds.), *Who learns what from cases and how?: The research base for teaching and learning with cases* (pp. ix-xv). Mahwah, NJ: Lawrence Erlbaum.

Merseth, K. K. (2003). *Windows on teaching math: Cases of middle and secondary classrooms.* New York: Teachers College Press. (Two books: case book and facilitator's guide)

National Council of Teachers of Mathematics. (1991). *Professional Standards for Teaching Mathematics.* Reston, VA: Author.

Patel, V. L., & Kaufman, D. R. (2001). The skilled craftsman and the scientist practitioner: Reflections on problem-based learning and medical education. *The Chronicle of Higher Education*, B12 Feb 2, 2001.

Richardson, V. (1996). Forward II. In J. Colbert, K. Trimble, & P. Desberg (Eds.), *The case for education: Contemporary approaches for using case methods* (pp. 197-217). Boston: Allyn & Bacon.

Schifter, D., Bastable, V., & Russell, S. J. (2007). *Patterns, functions, and change.* Parsippany, NJ: Dale Seymour Publications. (Two books: case book and facilitator's guide, video)

Seago, N., Mumme, J., & Branca, N. (2004). *Learning and teaching linear functions: Video cases for mathematics professional development, 6 – 10.* Portsmouth, NH: Heinemann.

Shulman, L. S. (1996). Just in case: Reflections on learning from experience. In J. Colbert, K. Trimble, & P. Desberg (Eds.), *The case for education: Contemporary approaches for using case methods* (pp. 197-217). Boston: Allyn & Bacon.

Shulman, L. S. (1986). Those who understand: Knowledge growth in teaching. *Educational Researcher, 15(2)*, 4-14.

Smith, M. S. (2001). *Practice-based professional development for teachers of mathematics.* Reston, VA: National Council of Teachers of Mathematics.

Smith, M. S., Silver, E. A., & Stein, M. K. (2005a). *Improving instruction in rational numbers and proportionality: Using cases to transform mathematics teaching and learning, Volume 1.* New York: Teachers College Press.

Smith, M. S., Silver, E. A., & Stein, M. K. (2005b). *Improving instruction in algebra: Using cases to transform mathematics teaching and learning, Volume 2.* New York: Teachers College Press.

Stein, M. K., Smith, M. S., Henningsen, M. A., & Silver, E. A. (2000). *Implementing standards-based mathematics instruction: A casebook for professional development.* New York: Teachers College Press.

Sykes, G., & Bird, T. (1992). Teacher education and the case idea. In G. Grant (Ed.), *Review of Research in Education* (Vol. 18, pp. 457-521). Washington, DC: American Educational Research Association.

---

[1]This chapter draws on ideas recently discussed in Markovits, Z., & Smith, M.S. (in press). Case as tools in mathematics teacher education. In D. Tirosh (Ed.), *Tools and processes in mathematics teacher education, The international handbook of mathematics teacher education, Volume 2.* Rotterdam, the Netherlands: Sense Publishers.

Margaret S. Smith is an Associate Professor in the Department of Instruction and Learning in the School of Education and a Research Scientist at the Learning Research and Development Center, both at the University of Pittsburgh. She works with preservice elementary, middle, and high school mathematics teachers at the University of Pittsburgh, with doctoral students in mathematics education who are interested in becoming teacher educators, and with practicing middle and high school mathematics teachers and coaches locally and nationally through several funded projects. Over the past decade she has been developing research-based materials for use in the professional development of mathematics teachers and studying what teachers learn from the professional development in which they engage. She is currently a member of the Board of Directors of the National Council of Teachers of Mathematics (2006-2009).

Susan N. Friel is a Professor of Mathematics Education in the School of Education at the University of North Carolina. She works with both preservice and inservice teachers, with a focus on K-8 mathematics teacher education. She is a frequent "case user" and has been part of pilot programs for the use of some of the case materials identified in this manuscript. Much of her work involves curriculum development, either for K-8 students or, as professional development materials, for K-8 teachers. She is co-author of the K-5 *Used Numbers* Program, of the 6-8 *Connected Mathematics* Program, and of the K-5 professional development program, *Teach-Stat.* Currently, her focus is on working with K-2 teachers, with an emphasis at grade 2, looking at the design of *purposeful pedagogy* which involves the interaction of problem-based learning with strategic interventions to enhance and support students' mathematics learning.

Morris, K. A.
AMTE Monograph 4
Cases in Mathematics Teacher Education: Tools for Developing Knowledge Needed for Teaching
©2008, pp. 9-20

2

What Would *You* Do Next if This Were *Your* Class? Using Cases to Engage Preservice Teachers in Their New Role

Katherine A. Morris
Sonoma State University

This chapter outlines a mathematics teacher educator's use of a narrative case in a preservice elementary methods course. Following a description of the instructional context, goals, sequencing, and pedagogical reasoning underlying the implementation of the case, this chapter focuses on preservice teachers' opportunities to learn from this particular case as it was implemented. Although there is clear evidence that the case enabled these preservice teachers to grapple with many important topics in mathematics education, they were not able to accomplish everything their instructor had hoped. This leads the author to raise important questions about what can be accomplished using cases and what counts as a successful case discussion.

"So, imagine these were *your* second grade students. And imagine this is the work they did when *you* gave them the lollipop problem. As their teacher, what would *you* do next?" Five weeks into the semester, this is the prompt that I used to initiate a whole-class discussion of the narrative case, "You just count the extras" (Barnett-Clarke & Ramirez, 2003, pp. 21-30) in my preservice elementary mathematics methods course. Within the case, when the teacher asks her students to solve a "how many more?" problem, she is surprised to find that many of her students are confused, despite direct instruction and practice the previous day on this type of problem.

It is well established that using cases can be an effective tool to help teachers see multiple perspectives and think through dilemmas of practice (e.g., Lundeberg, Levin, & Harrington, 1999; Shulman, J., 1992). By this point in the semester, the course texts[1] and in-class discussions had framed mathematics as a human construction that should be reasonable and sensible for both teachers and students; they had framed the role of the teacher in mathematics class as that of facilitator of student thinking (e.g., through problem selection, establishment of discourse norms, and use of representations). The preservice teachers had some experience observing and analyzing teaching practice, investigating curriculum, and evaluating student learning. They had just begun to plan and implement mini-lessons in their field placement. Although they had been exposed to a lot in a short period of time, it was not clear what they could do with what they were learning. I selected this particular case to simulate a scenario in which my preservice teachers needed to leverage previous class readings, discussions, and assignments to *do* something — to "think like a teacher" by evaluating their options, deciding what they would do next for these children if this were their class, and thinking through the possible implications of their decisions.

Background

Nearly all of the individuals in my elementary mathematics methods courses are enrolled in an intense one-year teaching credential program, and most concurrently take four 3-unit teaching methods courses (one in each of the core disciplines) and spend two full days each week in an elementary classroom. Primarily post-baccalaureate, middle-class females, ages 22-50, these preservice teachers vary widely in their mathematical skills and understandings, as well as in their dispositions. Sadly, only a few of these prospective teachers had taken our "mathematics for elementary teachers" content courses because they are no longer required to do so for the credential.[2] Hence, although this practice-oriented methods course focuses on pedagogy, curriculum, worthwhile tasks, student thinking, standards, dispositions for teaching mathematics, and the like, the class is also designed to cultivate a robust understanding of basic mathematics. Within the course, students routinely engage in and discuss tasks used in elementary classrooms to learn about children's development of mathematical ideas, instructional materials, and appropriate pedagogical strategies. A fundamental challenge in the methods course is to scaffold students' learning in such a way that, when they have their own classrooms, they will be able to engage in the kinds of teaching practices we have studied.

The previous week, these preservice teachers had worked with two categorization schemes from Cognitively Guided Instruction (CGI) (e.g., Carpenter, Fennema, Franke, Levi, & Empson, 1999; Carpenter, Fennema, Peterson, Chiang, & Loef, 1989). They were introduced to and created examples for CGI's four different kinds of addition/subtraction problems (i.e., *join*, *separate*, *part-part-whole*, and *compare*). Later, they analyzed video clips of children solving such problems and categorized both the problem type and the child's strategies (i.e., *direct modeling*, *counting strategies*, and *number fact strategies*). Although these preservice teachers could mechanically solve the basic subtraction word problems with ease, the idea that subtraction did not necessarily imply "take away" was novel. In class we spend a fair amount of time modeling and sketching representations for part-part-whole and compare problems. Figure 1 provides examples of the representations preservice teachers construct that reveal their new understandings of part-part-whole and compare problems.

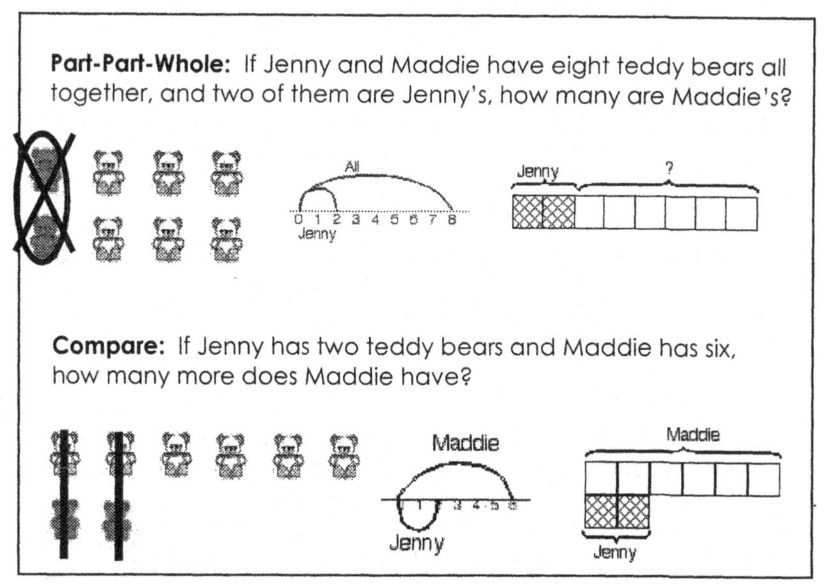

Figure 1. Examples of strategies used by students

Framing the Case

In general, the cases in Barnett-Clarke and Ramirez's book focus teachers' attention on what is confusing in primary mathematics and how aspects of language impact student learning, as well as on various teaching strategies that support conceptual development in young children. These attributes are attractive as they seem well suited to my goal to build on my preservice teachers' emerging curricular understanding and push them towards *practice*. I selected the "You just count the extras" case because it provided an apt practice-based follow-up to the theory I had introduced the previous week. In the case a second grade teacher, having been out for a few days, returns to find that the substitute has used the textbook to teach the class to use subtraction when a word problem asks them, "how many more?" Wanting to check for understanding, the teacher asks her class to solve the following problem. *Brandon had 14 lollipops. Jeffrey had only 6 lollipops. How many more lollipops did Brandon have?* (Barnett-Clarke & Ramirez, 2003, p. 21) The case goes on to describe how the second graders quickly got to work on the problem, some with and some without manipulatives, and how they recorded their work using pictures, words, and numerical expressions.

The case outlines the thinking of eight second graders and includes the written work (both correct and incorrect) of four students. Many second graders arrived at 20 as their solution, having added 14 + 6. Others modeled and/or drew Brandon's and Jeffrey's lollipops, matching up the groups and "counting the extras." These students arrived at eight as their solution. The brief case concludes by highlighting what the teacher wanted her students to understand and what she wondered about their thinking.

In addition to providing a realistic context in which preservice teachers can practice applying some of the theory they have learned, the case also served as a rhetorical device. I selected this case because I believed it would help me reinforce a set of interrelated notions of teaching and learning that had begun bubbling up in class discussions:

- children's mathematical understandings can be fragile; the goal is for them to become increasingly reliable, flexible, and general (Kilpatrick, Swafford, & Findell, 2001);
- just because a concept or skill was "taught" and "practiced" yesterday, does not mean students will "get it" today (Schifter & Fosnot, 1993);
- learning the language of mathematics is part of learning mathematics (e.g., Cobb, Wood, & Yackel, 1993; Kazemi, 1998);
- within most student errors, you can find a line of reasoning and understanding (e.g., Ball, Hill, & Bass, 2005; Kamii, 2000);
- tasks that bring misconceptions to the fore are precious intellectual resources teachers can learn to use (e.g., Fennema et al., 1996; Shannon, 1999), and such resources can be used publicly (e.g., Ball & Bass, 2000; Lampert, 2001); and
- when you find that students aren't "getting it," identify where the trouble is and, as the teacher, you have options, depending on your goals, what you think students *do* know, and what you want them to know (e.g., Heaton, 2000; Lampert, 2001).

Using the Case to Promote Teacher Thinking

One-third of the way into the semester, the use of this narrative case marked a juncture in the class. In addition to their emerging knowledge of the complexity of elementary mathematics and children's thinking about mathematics, my preservice teachers were beginning to articulate details about specific teacher "talk moves" that had been emphasized in course texts (e.g., Chapin, O'Connor, & Anderson, 2003). These included questioning strategies, the use of wait time, and strategic use of a variety of participation structures for specific interactions. Up to this point, however, when the class discussed teacher moves, the preservice teachers spoke mostly in the third person — discussing the moves of teachers whose practices they encountered through texts, and in both real and virtual observations. Thus, one of my instructional goals in using a case was to provide a context in which my students could use what they were beginning to learn about mathematics and student thinking to push their own thinking and teaching practice.

I wanted the case to provide my preservice teachers the opportunity to reason through dilemmas of teaching practice individually and collectively (Harrington, 1995; Lundeberg, Levin, & Harrington, 1999). Specifically, I wanted them to grapple with two related issues: what do you do when some of the students can solve a problem while others are struggling, and what exactly are the strugglers struggling with? Furthermore, I didn't just want them to think *about* teaching; I wanted to leverage this case to help the preservice teachers begin to think *like* a teacher.[3] I wanted them to imagine themselves as this teacher working with students' fragile understandings, uncovering students' misconceptions and points of confusion, and reasoning with a whole class of second graders about how to solve a comparison problem correctly.

The preservice teachers interacted with the case in three distinct phases. First, they worked individually with the case – reading it and taking notes on the mathematics, noting evidence of student thinking including possible points of confusion, and identifying ideas for how to proceed if it were their class. Then, consistent with Judith Shulman's (1992) assertion that dialogue is the key to learning from cases, the next two phases provided ample opportunities for discussion. In the second phase, preservice teachers worked in small groups to discuss and come to consensus about what they thought the students knew and could do, as well as what the "growing edges" might be (e.g., What were the second graders on the brink of understanding? What understandings seemed fragile or incomplete? What seemed confusing to them? What could they do only with scaffolding?).[4]

The preservice teachers were then asked to discuss what they would do next if this were their second grade class – what problems might they pose, what problem solving strategies might they emphasize, what models or representations might fit, and how might they scaffold the experience. While students talked in their small groups, I circulated, eavesdropping on each group's conversation, taking note of their interpretations of students' thinking and the issues they raised. Table 1 includes the four main domains for the small-group conversations. While all of the groups actively engaged in conversations about what students seemed to know and what seemed to be confusing/challenging for them (some insightfully so), not one of the groups focused on what they would do next if it were their class.

After about fifteen minutes, I interrupted the small-group discussions to reconvene as a whole class. With my goal firmly in mind, I introduced the third phase of our case work with the following prompt: "So, imagine these were *your* second grade students. And imagine this is the work they did when *you* gave them the lollipop problem. As their teacher, what would *you* do next?" Hands sprang up. One preservice teacher, Jim, quickly asserted that it was really an issue of language more than mathematics. Lanie suggested that there should be more real-life problems that children can relate to in order to help them reason through the scenario. The conversation took off from there; during the next 30 minutes of animated discussion nearly everyone in the class participated.

Table 1. Foci of Small Group Discussions

Language	**Students' Personal Connections**
• the word *difference* • the word *more*, what it signals and how it might be confusing in a subtraction problem • *more* vs. *fewer*	• the size of the numbers • lollipops vs. other objects • names of children in the story problem • students' reading abilities/abilities in English
Mathematics	**Manipulatives**
• What is subtracted in a comparison problem? • How do comparison problems connect to basic facts and fact families?	• the need for 20 items to model this comparison problem (14 – 6 = 8) • connections between direct modeling & paper/pencil representation of problem/solution • the choice of manipulatives

Unpacking the Case Discussion

Unlike many previous class discussions, the preservice teachers freely bounced ideas off one another, seldom relying upon me to either mediate or moderate the conversation. They proceeded to articulate their emergent understandings of the mathematics; their interpretations of student thinking about the lollipop problem; their interpretations of strategies for solving the problem; the myriad ways of modifying the problem to address points of confusion; and the tensions in constructivist teaching regarding when, why, and how to allow students to struggle with problems. Neither the discussion of strategies nor constructivist teaching tensions had been evident to me during the small-group discussions.

The questions and issues they raised for each other and discussed in earnest were both pertinent and interesting.

- Was it a good or bad problem if it was hard for students?
- Could second graders handle thinking about two kinds of subtraction problems at once, or would this just confuse them?
- If students had worked extensively with number lines as a model for solving addition and subtraction problems (as one preservice teacher had observed in her fieldwork), how might students use the number line to solve comparison problems?

- Should the teacher "go back" to more straightforward take-away subtraction problems for a while?
- How does a comparison problem fit with "fact families"? If children knew their facts, how would that help them with the problem?
- What if the manipulatives kids used were stick-shaped instead of squares and cubes? Would they use them differently?
- How is having two children acting out the word problem for the whole class like/unlike an individual student modeling the problem with manipulatives by him/herself? Does it matter?
- Is there an action analogous to "taking away" in a comparison problem?
- How much struggle is productive for students? What might the teacher do when children are stuck? Tell them to subtract? Ask them questions? Suggest easier numbers?
- What might children do when facing a comparison with much larger numbers?
- How do children generalize to see comparison problems as a class of problems that require subtraction? How does a teacher help students generalize? When would this be fitting?
- How does this fit with what we are supposed to be teaching according to the state standards?

At various points in the conversation I chose to weigh in, providing my perspective on specific issues/questions. I gave my opinion about whether or not this was a good problem for students of this age and about when I thought struggling could be fruitful. I added additional information about mathematical models, such as number lines and basic problem-solving strategies (e.g., act it out, model it, work backwards, use simpler numbers). I occasionally made explicit connections to previous course activities, assignments, and readings. But in general, the case discussion seemed to assume a life of its own.

My preservice teachers' animated tone, the degree of participation in a relatively unmediated conversation, their prolonged engagement in the issues suggested by the case, the number of connections they drew to ideas they had encountered elsewhere in the course (and their credential program), and the sheer number of ideas about teaching mathematics they grappled with simultaneously suggests that this case, as an instructional tool, was successfully employed. That is to say, there is ample discursive evidence that the case discussion provided my preservice teachers "opportunities for learning" (SBCDG, 1993; Tuyay, Jennings, & Dixon, 1995). Following the case activities, "You just count the extras" became a touchstone — an intertextual reference (SBCDG, 1992) — for some of the preservice teachers, serving as a reference point that periodically emerged in subsequent discussions of word problems, issues of language, modeling, and the like.

Nevertheless, despite my repeated prodding, "So what would *you* do if it were *your* class?", the preservice teachers would not take the bait. They firmly maintained the stance that had previously been developed in my course and throughout the program: raising questions and wondering rather than making assertions and reasoning through a position. Although using the case was successful on some levels, it clearly did not accomplish all of my intended purposes. On the one hand, using the case in conjunction with theory and standards provided a rich context that stimulated engaging, productive, prolonged practice-based discussions grounded in issues from readings and previous course activities. On the other hand, the case discussions also highlighted for me, as the instructor, that these preservice teachers seemed to be unwilling or unable at this point in the semester to try the role of teacher, even in a clearly hypothetical situation where they could play around with what they *might* do if this *were* their class.

What is Success in Implementing Instructional Cases?

I had hoped that the case might enable me to create an intellectually honest yet safe space in which the preservice teachers could collaboratively posit potential next steps for the second grade class as the teacher, and then sustain a conversation in which they thought through various implications. In essence, I hoped to use the case as the basis for a series of "thought experiments" to simulate both day-to-day and in-the-moment instructional planning in which the teacher is at once responsive to his/her students and grounded in both theory and standards. But the preservice teachers would not play along. Thus I was unable to leverage the case to help them practice "thinking like a teacher." Clearly they had no inhibitions about discussing the case, but they would not locate themselves within the scenario. They resisted putting themselves in a situation where *they* needed to determine a course of action, despite the ambiguities and dilemmas inherent in the situation (Harrington, 1999). They could address the possibilities and dilemmas in the abstract ("the teacher could…") but they could not themselves assume the role of teacher ("if it were my class, I would…").

The preservice teachers' reluctance leads to the questions I now pose for myself as I begin to plan for the coming semester, but also more broadly for the community of mathematics teacher educators. Given that the case was substantive enough to engage my students' interest for a prolonged period of time, was the case discussion successful despite the fact the preservice teachers were unable to assume the role of teacher in this instance? What were the factors that precluded them from so doing? Were there factors in the way the narrative case was constructed, in the portrayal of context, students, teacher or curriculum, that did not ring true enough to permit preservice teachers to see themselves in the role? Was there something in the enactment such that they did not buy into the idea of the thought experiment? Were they actively resisting assuming the teacher's role? Or, rather, were there "developmental" issues that kept the preservice teachers from being able to perceive themselves as sufficiently prepared to assume the role of mathematics teachers at this early point in the semester? From my vantage point, it is difficult to tell, and in truth I suspect there is a complex web of factors contributing to their responses.

Nevertheless, such questions seem relevant, particularly in light of the ever-expanding range of high-quality case-based instructional materials available in mathematics teacher education. For with these materials comes the need for mathematics teacher educators to develop a more refined "pedagogy of cases" (Shulman, L., 1992). Not only do we need to develop a more nuanced understanding of what cases, or particular kinds of cases, can and cannot do, we also need to develop more nuanced understandings of what different kinds of teacher learners (e.g., preservice, novice, veteran, teacher leader) can learn from and/or do with different kinds of cases. More to the point, we also need new tools that are fine-grained and poly-focal to determine what counts as successful implementation of cases in mathematics teacher education. Is it enough for a case to lead to an interesting conversation about worthwhile topics in mathematics education? Or are there specific learning outcomes or teaching practices that must be evident for the case enactment to be successful? What should count as a warrant for the claim that the use of a case was successful? It is time we move from our general hunch that using cases in mathematics teacher education is good practice and move toward a robust understanding of just what the practice of using cases can (and perhaps cannot) accomplish and a better means of determining the efficacy of using cases to teach mathematics teachers.

Given all these questions, what do I think about the effectiveness of my implementation of Barnett-Clarke & Ramirez's case? Did I successfully deploy it and did my preservice teachers learn from the experiences? For me the answer is a qualified "yes." Though unable to take-on the role of the teacher at this early point in the course, throughout the case discussion, the preservice teachers' interpretations of student misconceptions, use of appropriate mathematical models, and suggestions for related problems demonstrated a depth in their mathematical knowledge for teaching (Ball, Hill, & Bass, 2005). Perhaps that is an appropriate instructional objective for an activity that comes early in the semester. Knowledge development, however, is a necessary but insufficient factor in becoming an effective math teacher, because teaching is a practice and one must be able to make reasoned decisions on the fly and ultimately one must act. The "You just count the extras" case got them started on the path – but I wanted them to go farther!

Barnett-Clarke & Ramirez (2003) wrote their casebook with practicing teachers in mind. They write, "As teachers we make thousands of decisions daily. Few of them are clear-cut; usually, they require trade-offs" (p. 2). It is precisely this kind of teacher thinking that my preservice teachers could not manage. Perhaps this is in part because, unlike their inservice counterparts, early in their certification program they simply do not have sufficient experience to conjure a robust image of a second grade math lesson, much less see themselves in the role of decision-maker. I have not abandoned the use of this case, but I have modified my expectations for what it will help preservice teachers accomplish.

But I still want to see how preservice teachers think on their feet in the context of a math lesson. And so, based on my findings from the case discussion reported herein, I am experimenting with another pedagogical strategy to help preservice teachers simulate "what would you do next" decision-making. Preliminary results indicate that using a *video* case at the end of the course is providing the preservice teachers the verisimilitude that they need to step into role of classroom teacher. At pre-selected junctures in the video I pause the video and ask my class — "And what would *you* do next if this were *your* class?"

References

Ball, D. L., & Bass, H. (2000). Making believe: The collective construction of public mathematical knowledge in the elementary classroom. In D. Phillips (Ed.), *Yearbook of the National Society for the Study of Education, Constructivism in Education*, (pp. 193-224). Chicago: University of Chicago Press.

Ball, D. L., Hill, H. C, & Bass, H. (2005). Knowing mathematics for teaching: Who knows mathematics well enough to teach third grade, and how can we decide? *American Educator*, February, 14-17, 20-23, 43-46.

Barnett-Clarke, C., & Ramirez, A. (Eds.). (2003). *Number sense and operations in the primary grades: Hard to teach and hard to learn?* Portsmouth, NH: Heinemann.

Carpenter, T. P., Fennema, E., Franke, M. L., Levi, L. & Empson, S. (1999). *Children's mathematics: Cognitively Guided Instruction.* Portsmouth, NH: Heinemann.

Carpenter, T. P., Fennema, E., Peterson, P. L., Chiang, C. P., & Loef, M. (1989). Using knowledge of children's mathematical thinking in classroom teaching: An experimental study. *American Educational Research Journal, 26* (4), 499-531.

Chapin, S., O'Connor, C., & Anderson, N. (2003). *Classroom discussions: Using math talk to help students learn, Grades 1-6.* Sausalito, CA: Math Solutions Publications.

Cobb, P., Wood, T., & Yackel, E. (1993). Discourse, mathematical thinking, and classroom practice. In E. Forman, N. Minick, and C. A. Stone (Eds.), *Contexts for learning: Sociocultural dynamics in children's development* (pp. 91-119). New York: Oxford University Press.

Fennema, E., Carpenter, T. P., Franke, M. L., Levi, L., Jacobs, V., & Empson, B. (1996). A longitudinal study of learning to use children's thinking in mathematics instruction. *Journal for Research in Mathematics Education, 27*(4), 403-434.

Fenstermacher, G. D. (1994). The place of practical arguments in the education of teachers. In V. Richardson (Ed.), *Teacher change and the staff development process: A case in reading instruction* (pp. 23-42). New York: Teachers College Press.

Fosnot, C., & Dolk, M. (2001). *Young mathematicians at work: Constructing multiplication and division.* Portsmouth, NH: Heinemann.

Harrington, H. L. (1995). Fostering reasoned decisions: case-based pedagogy and the professional development of teachers. *Teaching and Teacher Education, 11*(3), 203-221.

Harrington, H. L. (1999). Case analysis as a performance of thought. In M. A. Lundeberg, B. B. Levin, & H. L. Harrington (Eds.), *Who learns what from cases and how?: The research base for teaching and learning with cases* (pp. 29-48). Mahwah, NJ: Lawrence Erlbaum.

Heaton, R. (2000). *Teaching mathematics to the new standards: Relearning the dance.* New York: Teachers College Press.

Kamii, C. with Housman L. B. (2000). *Young children reinvent arithmetic: Implications of Piaget's theory. 2nd ed.* New York: Teachers College Press.

Kazemi, E. (1998). Discourse that promotes conceptual understanding. *Teaching Children Mathematics, 4*(7), 410-414.

Kilpatrick J., Swafford, J., & Findell, B. (Eds.), (2001). *Adding it up: Helping children learn mathematics.* Washington, DC: National Academy Press.

Lampert, M. (2001). *Teaching problems and the problems of teaching.* New Haven, CT: Yale University Press.

Lundeberg, M. A., & Fawver, J. (1994). Thinking like a teacher: Encouraging cognitive growth through case analysis. *Journal of Teacher Education, 45*(4), 289-297.

Lundeberg, M. A., Levin, B. B., & Harrington, H. L. (Eds.). (1999). *Who learns what from cases and how?: The research base for teaching and learning with cases.* Mahwah, NJ: Lawrence Erlbaum.

Santa Barbara Classroom Discourse Group (SBCDG). (1992). Do you see what we see? The referential and intertextual nature of classroom life. *Journal of Classroom Interaction, 27*(2), 29-36.

Santa Barbara Classroom Discourse Group (SBCDG). (1993). Talking knowledge into being: Discursive and social practices in classrooms. *Linguistics and Education, 5*(3-4), 231-239.

Schifter, D., & Fosnot, C. (1993). *Reconstructing mathematics education: Stories of teachers meeting the challenge of reform.* New York: Teachers College Press.

Shannon, A. (1999). *Keeping score.* Washington, DC: National Academy Press.

Shulman, J. H. (Ed.). (1992). *Case methods in teacher education.* New York: Teachers College Press.

Shulman, L. (1992). Toward a pedagogy of cases. In J. H. Shulman (Ed.), *Case methods in teacher education* (pp. 1-30). New York: Teachers College Press.

Trafton, P., & Thiessen, D. (1999). *Learning through problems: Number sense and computational strategies.* Portsmouth, NH: Heinemann.

Tuyay, S., Jennings, L., & Dixon, C. (1995). Classroom discourse and opportunities to learn: An ethnographic study of knowledge construction in a bilingual third grade classroom. *Discourse Processes, 19*(1), 75-110.

---

[1] For this class, rather than reading a single textbook on elementary mathematics teaching methods, preservice teachers read a number of trade books, including Trafton & Thiessen (1999) which focuses on teaching practices that support big problems with small children; Fosnot & Dolk (2001) which focuses on how to develop robust understandings of multiplication and division by focusing on children's concepts, strategies, and models; and Chapin, O'Connor, & Anderson (2003) which focuses on specific strategies for fostering productive mathematical discourse. Five weeks into the semester, they have nearly finished the first book and have read a few chapters in each of the other two.

[2] In fact, in California there seems to be a disturbing decrease in the percentage of prospective elementary teachers in these content courses. Anecdotal reports indicate this is a direct result of the state-mandated exam that has supplanted content courses and has become the only means by which students can demonstrate "subject-matter competence" as a "highly qualified teacher" in compliance with NCLB.

[3] Although Lundeberg & Fawver (1994) use the term "thinking like a teacher" in their work in which teachers formally analyzed dilemma-based narrative cases, here I take my cue from Fenstermacher (1994), intending something more akin to teacher's "practical reasoning." That is to say, I want the preservice teachers to identify the premises that link a teacher's reasoning to her decisions/actions.

[4] The term "growing edges" was introduced to me by Gena Richman, a mentor teacher who regularly works with student teachers from my university and is a valued co-researcher. For me this metaphor conjures images of growth in nature (e.g., leaves, crystals), and I connect it with Vygotskian notions of zone of proximal development.

Katherine A. Morris is an Assistant Professor at Sonoma State University. As a Goldman-Carnegie QUEST Fellow (Carnegie Foundation for the Advancement of Teaching), she explored the use of cases as a potential signature pedagogy in teacher education. As part of that work, she created a web-based record of her own teaching practice focused on use of K-12 multimedia records of practice in her methods course. You can view this project at http://gallery.carnegiefoundation.org/insideteaching. She is also investigating how a constructivist third grade teacher uses an instructional routine to foster mathematical understandings and productive dispositions through discourse and representation.

Kazemi, E., Lenges, A., and Stimpson, V.
AMTE Monograph 4
Cases in Mathematics Teacher Education: Tools for Developing Knowledge Needed for Teaching
©2008, pp. 21-33

3

Adapting Cases from a *Developing Mathematical Ideas* Seminar to Examine the Work of Teaching Closely

Elham Kazemi
University of Washington

Anita Lenges
The Evergreen State College

Virginia Stimpson
University of Washington

This chapter describes three uses of cases from the Building a System of Tens (BST) seminar. BST is one of seven modules available in the Developing Mathematical Ideas professional education materials (Schifter, Bastable, & Russell, 1999). BST is designed to support elementary teachers in analyzing how students develop a robust understanding of the base-ten system. The cases help seminar teachers learn how students' thinking develops when they are given opportunities to share and explore their understanding and their confusions. We describe how we have used particular cases to achieve a focused set of goals especially targeted for teachers who are beginning to learn to build instruction on students' thinking. We explain how our strategic use of cases brings to the surface the significant work that teachers do when they design tasks and anticipate, elicit, and respond to students' mathematical ideas.

This chapter describes three uses of cases from the *Building a System of Tens* (BST) seminar. BST is one of seven modules available in the *Developing Mathematical Ideas* professional education materials (Schifter, Bastable, & Russell, 1999). BST is designed to support elementary teachers in analyzing how students develop a robust understanding of the base-ten system. The cases help teachers become familiar with strategies students use as they solve whole number computation problems across the operations. Over the course of the seminar, teachers do mathematics together, study written and video cases of students in classrooms, and share episodes from their own classrooms as they try new ideas with their own students. Through participating in DMI seminars, teachers also consider new pedagogical practices inspired by seeing the sense in students' work. The seminars are meant to help teachers appreciate how students' thinking develops when they are given opportunities to share and explore their understanding and their confusions. Teachers examine the logic in students' mathematical strategies and learn ways to challenge and extend their understanding.

In this chapter, we each describe how we have used particular DMI cases to achieve a focused set of goals especially targeted for teachers who are beginning to learn how to elicit and use children's thinking to guide instruction, whether as preservice teachers or as new or seasoned practicing teachers. Our expanded use of these cases grew out of our collaborative engagement in long-term professional development, including extensive experience facilitating DMI seminars.[1] The activities and questions in *Building a System of Tens* tend to focus primarily on understanding the mathematical concepts and ideas that underlie students' mathematical thinking. Because of our familiarity with the seminars, we recognized that particular cases could be adapted strategically to further teachers' engagement with the work they do when they anticipate, elicit, and respond to students' mathematical ideas. These three adaptations do not necessarily need to follow each other in a professional development course. However, if these adaptations are used within the context of a DMI seminar or similar contexts, then we believe that together they will help teachers pay more attention to their role in using children's thinking to guide instructional decisions. We have used these adaptations in a variety of different settings, including DMI seminars, preservice methods courses, week-long summer institutes, and inservice professional development days. The adaptations reflect our orientation to make explicit what teachers do as they interact with subject matter, tasks, and children's mathematical ideas in the classroom.

Cases are an increasingly popular method to support teacher learning across preservice and inservice education. Much has been written about how cases of classroom instruction are particularly well-suited to capture the complexities of teaching and how they can promote critical inquiry among teachers (Broudy, 1990; Harrington & Garrison, 1992). There is a clear call in the literature to specify what and how teachers learn from cases (e.g., Lundeberg, Levin, & Harrington, 1999; Sykes & Bird, 1992).

The cases summarized in this chapter all highlight students' mathematical thinking. Our view is that teachers' efforts to understand student thinking can deepen their own disciplinary knowledge and should guide their consideration of how to pose mathematical problems and facilitate mathematical work in the context of their classroom. We have adapted these specific DMI cases to highlight the instructional thinking and decisions that teachers make as they engage with their students' ideas and decide on the tasks to use with students. This aspect of teachers' work can often be vague and unwieldy. Our adaptations are designed to make teaching transparent for the teacher who is at the initial stages of learning how to center instruction on children's thinking (see also, Stein, Engle, Hughes & Smith, in press).

Adaptation One: Pairing Cases 13 and 28 to Develop Place Value Understanding and Reason about the Design of Classroom Tasks

In what follows, Stimpson begins by describing how she has used two cases that appear in two different sessions of a typical DMI seminar to help teachers connect ideas about place value in whole and decimal numbers. This illustrates one way of leveraging the content of particular cases to further teachers' mathematical understanding and help them think about how to use that knowledge when they design tasks in the classroom.

Summarizing the Cases

Case 13, *Ones, Tens, Hundreds,* focuses on the substantial work students must do to make sense of the spoken and written number systems and relationships between the two systems. Marie, a third-grade teacher, describes a lesson that was designed to revisit the writing of numbers, focusing on the value of the digits in the hundreds, tens, and ones places by using expanded notation and representing these quantities with base-ten blocks and pictures.

The teacher describes the variety of ways that students choose to represent three-digit numbers. As an example, several students mistakenly represent 426 with 4 flats, 6 rods, and 2 units, as shown in Figure 1. The case includes dialogue that shows how the teacher probes to learn more about one student's thinking behind this representation. The teacher does not step in and correct the student's representation. By asking questions to learn about the student's thinking, she provides people who read her case with an opportunity to consider the logic in the student's work and what more there is for this student to learn.

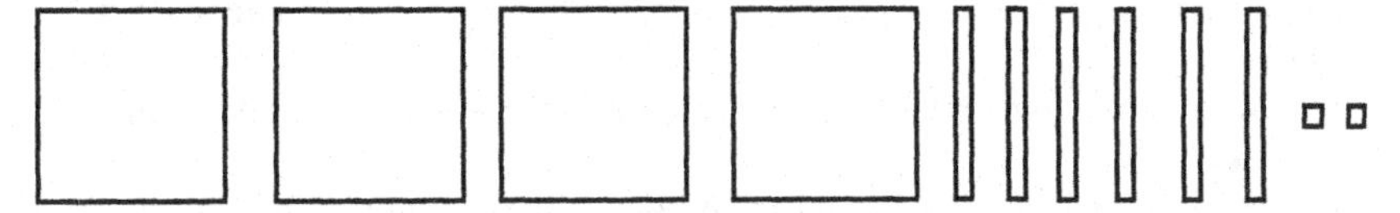

Figure 1. Students' mistaken representation of 426 with Base Ten Blocks

This first case is paired with Case 28, *When the Answer is Important,* which explores ideas about the base-ten system that are revisited or revised when students explore the addition of decimal numbers. A fifth-grade teacher, Nicole, asks her students to return to their study of decimals after a break of several weeks. She offers a problem that provides an opportunity for students to consider how to add jagged decimals.

> "Pretend you are a jeweler," I said. "Sometimes people come in to get rings resized. When you cut down a ring to make it smaller, you keep the small portion of gold in exchange for the work you have done. Recently you have collected these amounts." I wrote on the board: 1.14 g .089 g .3 g "Now you have a repair job to do for which you need some gold. You are wondering if you have enough. Work together with your group to figure

out how much gold you have collected. Be prepared to show the class your solution" (p. 114).

After posing the problem and launching the students into small-group work, the teacher moves about the classroom, looking for opportunities to probe and extend students' understanding. The case includes several exchanges that reveal students' insights and confusions about adding decimal numbers. There is also a detailed description of whole-group sharing. Many students write in their math journals that a particular demonstration using a large cube to represent one, the flat to represent tenths, the long to represent hundredths, and the small cube to represent thousandths made it clear that you need to add tenths to tenths and hundredths to hundredths or you wouldn't get the "right" answer.

Using the Cases to Promote Place Value Understanding and Task Design

I initially paired these two cases when DMI facilitators reported they were losing the attention and participation of some primary teachers once they hit the session that focused on decimals. I wondered how we might help participants explicitly connect ideas across grade levels and consider features of representations that capture the relationship between adjacent digits in numbers, whether they involve decimals or not. It seemed the teachers needed opportunities to develop a robust understanding of the base-ten system by developing and *justifying* multiple representations for whole and decimal numbers.

I used the following questions to focus small-group conversation on the pair of cases.

1. Consider Case 13, *Ones, Tens, Hundreds,* written by a third grade teacher, Marie.
   - What is the logic behind the representations used by the students in this case?
   - What questions would you like to pose to these students in order to learn more about their understanding of place value?
   - What more about place value and representations is there for them to learn?
   - If you were their teacher, what task(s) or question(s) might you pose in order to extend the students' thinking?
2. Turn to Case 28, *When the Answer is Important,* by a fifth grade teacher, Nicole.
   - What did the students say or do that raised your curiosity or surprised you?
   - By the end of the case, what do some students understand?
   - What did the teacher do to help her students increase their understanding of place value?
   - What questions or tasks would you like to pose to these students in order to learn more about their understanding of place value?

What Teachers Learn About Place Value Understanding and Choosing Tasks

Most teachers became quite engaged in analyzing the third-grade students' thinking. Relatively quickly they recognized that students focused on single digits but not necessarily on the relationship between digits. That is, students were not showing they understood that each number to the left represents a multiple of a higher power of ten. At times I needed to press participants to find a possible logic for using 4 flats, 6 rods, and 2 units to represent the number 426 (see Figure 1). One conjecture offered was that, because 2 is smaller than 6, the student uses a smaller object to represent that digit. Others argued against that conjecture because the student used 4 flats to represent the 400. An important next move was to ask participants to suggest numbers they would ask the student to represent in order to test their various conjectures. As examples, some suggested 422 to learn whether or not the student would use similar-sized pieces for the 2 twenties and 2 ones. Others suggested 462 to learn if the student would now use 2 rods and 6 units. This work helped the teachers consider the logic in students' correct and incorrect answers and the importance of finding ways to verify conjectures about student thinking.

Often teachers puzzled about what task(s) or question(s) to pose in order to challenge and extend students' thinking. Some opted for reiterating which manipulative to use for 1s, 10s, 100s, etc. Others argued that there is something teachers understand about why base-ten blocks can be useful in representing place value that the students are missing. To deepen their understanding of powerful representations for numbers that spanned whole numbers and decimals, students might be invited to design their own representations for numbers as the students did in the second case.

Another suggestion was to highlight that some students represented 426 with 4 flats, 6 rods, and 2 units while others used 4 flats, 2 rods, and 6 units. If the class was going to agree to use only one of these representations, ask students if there is a justification for choosing one over the other? In other words, does one representation better represent 426 than the other? If yes, why?

The small-group and whole-group discussions about the second case were equally engaging. Teachers reported their surprise to read that students are not clear about why they line up decimals the way they do in order to add them. Teachers paused to think about how they line up whole numbers versus decimals and then began to talk about focusing on the place value rather than whether to line the numbers up on the right or line up the decimal points. This shift in focus helped participants recognize there are more similarities than differences when adding these two types of numbers.

Conversations about the two cases caused teachers to begin to question their own understanding of place value and to become curious about their own students' thinking. Some continued to suggest that if students were just given the right manipulative or more clearly stated rules then they wouldn't have these problems. Most moved to consider the powerful understandings students might develop if they are given an opportunity to reveal their current understandings and have experiences with representations and contexts that engage them in grappling with problematic or limiting aspects of their current thinking.

Adaptation Two: Using Case 6 to Focus on Teacher Questioning

Below, Lenges describes the use of Case 6, *Keeping it Straight,* to examine how a teacher carefully attends to what students are doing and asks purposeful questions to help students' advance their own understandings.

Summarizing the Case

In this case Lucy, the teacher, describes how her third grade student, Sarah, solves the problem 39 + 52. Although she can solve the problem correctly, she tries to come up with a new and unusual way to solve the problem so she can share it with the class. She chose to work with Unifix cubes, using yellow cubes for tens and black cubes for ones. Sarah got out 3 yellow cubes and 9 black cubes for the 39, and 5 yellow and 2 black cubes for the 52. She stacked together the 3 and 5 yellow cubes, and the 9 and 2 black cubes. She saw that there were 11 black cubes so she broke off 10 of them and joined them with the 8 yellows. Now she had 8 yellow and 10 black cubes in one stack, and one black cube in another stack. She was confused. She knew the answer was 91 but the cubes didn't look like 91. Previously, when solving problems like this using the traditional algorithm, she talked about "carrying the 1." The teacher was puzzled about what Sarah was thinking when she brought 10 cubes over to the 10s rather than exchanging the 10 ones for 1 tens cube.

The rest of the case includes the dialogue between Lucy, the teacher, and Sarah as Sarah solves a similar problem, 45 + 39. Lucy carefully listens to and watches Sarah, asks her clarifying questions, and never leads Sarah to think about the problem in a particular way. Sarah eventually has an "ah-hah" moment when she recognizes that 10 ones is the same as *a* ten, not 10 tens. Lucy urged Sarah to clarify and confirm her own thinking about the problem and the contradiction in answers. Sarah related her work back to the traditional algorithm and became very satisfied with her insights.

Using the Case to Highlight Teacher Questioning

This case is a useful example of how a teacher can support a student to make sense of an important idea by closely following the student's ideas and asking clarifying questions. It is challenging for teachers to see unfamiliar instructional strategies that help students construct their own understanding of mathematics. This case is simpler than others because it involves dialogue between one teacher and one student, avoiding the greater complexities of following a whole class dialogue.

In order to help make Lucy's strategies explicit and tie them to Sarah's learning, I start by asking teachers to read the case. After reading the case, in groups of three, with Unifix cubes on the table, I ask the groups to act out the exchange. One person acts as Lucy, a second person as Sarah, and the third is the narrator (there are a few places in the vignette where a narrator speaks as well). It may seem excessively explicit to ask teachers to read the case and then actually act out the dialogue and use the manipulatives, but after doing this with many groups it is clear that reading alone is insufficient to help teachers see what Sarah is doing and learning and how the teacher is interacting with her. When teachers are learning about student reasoning, they have to think carefully about what it is like for students to develop an understanding of something we take for granted—that 10 ones equals one ten—and the consequences of that knowledge. Acting out the dialogue also allows a teacher to *try on* new words and ways of interacting with a student.

After acting out the vignette, I ask each teacher to work alone again, to examine what the student learned and what the teacher did to support the student's conceptual development. The handout they work on is adapted from a session in which I participated led by Margaret Smith and included in Appendix C of Smith, Silver, and Stein (2005), *Improving instruction in Algebra: Using cases to transform mathematics teaching and learning, Volume 2.* I ask teachers to identify what Sarah understood before she began the dialogue with Lucy provided in the case. It is important for the teachers to be clear about what Sarah already knew. It will help them to be clear about what she was learning in the case. Some participants write that Sarah had memorized procedures for solving addition with regrouping but didn't really understand regrouping. Other teachers write that Sarah understood regrouping but was just confused by the manipulatives. After articulating what they believe Sarah already knew, I ask teachers to identify places throughout the vignette where Sarah is in the process of learning something. I label one column in a handout, "What Sarah learned or was in the process of learning," and the second column, "What Lucy did to facilitate or support Sarah's learning." By tying student learning to teacher moves, teachers can potentially develop a stronger understanding of teaching and questioning as it relates to learning rather than covering material. The third column asks teachers to cite a line number, provided in DMI case materials, to identify evidence for his/her claims.

After teachers have had about 20 minutes to work on this table, I pull the group together for a whole group discussion. I ask participants to share what Sarah learned or was in the process of learning. This may be skills, knowledge, or dispositions. Although there were initial differences in what they thought she already knew, the group generally sees how Sarah is developing stronger or more nuanced understanding of regrouping. As each participant shares her ideas about what Sarah was learning, I ask the participant to cite the places in the text supporting that idea. This is an important pedagogical move. We often make assumptions that our students are learning particular things because we hope they are, or because they have an opportunity to learn them. Or, teachers participating in a case discussion can often become critical of the teacher featured in the case when the case teacher allows students to struggle with ideas. But as we work hard to look closely at students' thinking and what they are actually learning, we need to attend to specific evidence of student learning. For example, when working through this case, many teachers suggest that Lucy should have provided Sarah with different manipulatives, such as Base 10 blocks, so she would not have been confused. It is not easy for some teachers who read this case to recognize the value in Sarah's struggle and to realize what she needed to make sense of and came to terms with because of her struggle.

In sharing what Sarah was in the process of learning, teachers often suggest that she is learning how to use Unifix cubes, or learning about regrouping, persistence, or the relationship between the traditional algorithm and Unifix cubes. Once teachers have identified several things that Sarah may have learned or been in the process of learning, I choose to highlight how Sarah develops a clearer understanding of the relationship between ones and tens and the idea of regrouping. Our number system does not allow you to note the value of 10 in the ones column, even though Sarah can create 11 black cubes, representing 11 "ones" with Unifix cubes. I do not take up the idea of how Sarah is learning about how to use Unifix cubes, because it is not a key mathematical idea. By focusing on Sarah learning about regrouping, we can track several teacher moves that support her learning. Using regrouping as a focus, I ask the teachers, "What did Lucy do to help Sarah learn about regrouping and the relationship between one ten and ten ones?" Teachers have already thought a lot about this as they worked individually. As teachers share their insights, I continue to probe for evidence, "Where did you see Lucy do that?" I also urge teachers to challenge each others' ideas, "Do you agree with what she just said?" As the conversation proceeds, I take notes on an overhead transparency of the teacher moves and line numbers that supported Sarah's learning. It is sometimes hard to hold the conversation to the case of Lucy and Sarah. Participants want to relate what is going on in the case to their own classroom. There is a place and time for this, but I try to stick to the case to develop collective clarity on just what happened in the case. Once we have identified key questions Lucy posed to Sarah around one aspect of Sarah's learning, I ask the teachers, "What was surprising to you as you read this case? What ideas does this give you for your own classroom?" Now teachers can make those connections and reflect on their own practice.

What Teachers Learn through Careful Study of Questioning Strategies

I have seen teachers gain several key insights by moving through this case in this way. Several teachers have been surprised that a student who could effectively solve two-digit addition with regrouping using the standard algorithm could be confused when using Unifix cubes. It raised a question for them: "When I knew a student could accurately solve a problem I assumed it meant that they understood what was going on. I'm going to ask my students to solve the problem with Unifix cubes to see what happens." Some teachers have thought that the manipulatives just confused Sarah and she just needed to learn how to regroup. Sarah's confusion didn't much trouble these teachers. As we discussed teacher moves, several teachers have been struck by Lucy's questions and her patience in helping Sarah develop an important understanding. They were encouraged to consider asking their students more clarifying and probing questions and to follow their students' ideas more closely. They note that as a teacher, "It is hard for me not to just point out small mistakes a student is making." Slowing down was a theme for more teachers; they thought that a bit more time with individual students, even those who seem to "get it," may uncover important areas where they need to learn. Finally, Lucy was persistent in having Sarah articulate what she learned. So even at a point where most teachers would be satisfied that the student "got it," Lucy made a comment, "I'm not sure I know what you mean." Though as a teacher we suspect Lucy did understand, she encouraged Sarah to clarify and articulate for herself her new mathematical insight.

There are two key issues I would like to highlight about the way I have used this case. First, I emphasize the importance of connecting teacher moves with student learning. While in professional development, we often work in the abstract on topics such as questioning strategies, mathematical ideas, and problem posing; effective teaching strategies should follow from students' ideas. Second, this case in particular allows us to focus on one student's learning. Although it is useful to tie student learning to teacher moves in a whole-class discussion, when teaching it is much more difficult to track individual student learning within a whole group discussion. Teachers often make the mistake of accepting a series of single-student responses to questions as indicative of collective understanding. By closely examining the dialogue between one student and one teacher, we can see the complexity of individual student learning.

Adaptation Three: Using Case 19 to Anticipate and Respond to Students' Thinking

Next, Kazemi describes how she adapted a DMI case to help teachers learn that anticipating student strategies and planning for a range of understandings that a mathematical task generates are core aspects of teachers' work. Case 19 in the BST casebook is well-suited to such an analysis because the teacher describes both the kinds of confusions and understandings students experienced as they worked with the task and wondered how to respond.

Summarizing the Case

This case describes fourth graders working on a lesson from the *Investigations* curriculum (Economopoulos, Russell, & Tierney, 1998) in which students solve harder two-digit multiplication problems by using what they know about easier ones. The case begins with a problem the students are exploring. In the *Investigations* curriculum, these are called cluster problems. For example, in order to solve 23×4, they first solve a set of easier problems, such as 2×4, 3×4, 2×40, 20×4, and then determine which of those can help them solve 23×4. In effect, students are deciding which problems to *cluster* in order to help them solve the harder problem. The sequence of problems is designed for students to develop efficient means to multiply by taking advantage of how to decompose harder multiplication problems into easier-to-recall subproducts. The teacher explains that her students typically will use 20×4 and then add on three more fours to get 92. As they work on these types of problems, she has found that most students have become fairly comfortable with the task and dive in eagerly. The majority of the case then focuses on a day when she asks students to generate their own cluster problems. After trying one together as a class, students work individually while the teacher sits with a group of eight students who asked for help. We learn about students' confusions, which we describe below, as the teacher recounts her work with this smaller group. Then, when she checks with other students in the class, we hear about what students who were eagerly working on the task encounter. The students who are most comfortable with cluster problems compete to use the biggest numbers possible and want to generate their own problems with large numbers.

This case is set within a DMI session in which teachers explore the meaning of multiplication, develop array models to explore the distributive property, and analyze why strategies that work for addition do not generalize to multiplication. The case discussion questions for the session focus teachers' attention on the meaning of multiplication as an operation. I noticed that I could use this case to also help teachers think about how to anticipate and respond to students' confusions and understandings.

Using the Case to Anticipate and Respond to Student Thinking

I adapted and extended the discussion of this case in the following way. We begin by reading the relevant pages of the fourth grade *Investigations* curriculum so that teachers have an opportunity to read the actual curriculum materials, become familiar with the guidance and information they provide, and understand what cluster problems are. I focus this reading with the following questions:

1. What is the mathematics that cluster tasks are supposed to elicit?
    a. What are cluster problems? Try a few.
    b. How would you describe the purpose of having students work on them?
2. What guidance does the curriculum give you for designing the lessons?

Taking the time to read the materials allows everyone, regardless of grade level, teaching experience, and familiarity with *Investigations*, to discuss the supports the materials provide.

After teachers read the curriculum materials and try the cluster task themselves, we move to discussion of the case. I pose the following three questions.

1. Based on case 19, what are some potential confusions that students might have as they work on these cluster problems (consider what the teacher says beginning in line 150, line 180, and line 195)?
2. How would you provide support for children who get confused by cluster problems? What kinds of modifications would you make?
3. How would you provide support for children who are comfortable with cluster problems? (Consider what the teacher says in line 125, line 155, and line 175.)

These questions are designed to encourage teachers to look closely in the case for evidence of the range of experiences that students have with this problem. Suggesting line numbers is meant to direct teachers to specific details about what makes this task challenging for some and transparent for others.

When we move to whole group discussions about students' confusions and understanding, I organize the discussion given the time constraints I have. Usually I begin by listing how participants describe the range of confusions students experienced. Then, I assign each of the confusions we listed to one or two small groups. They spend 15 to 20 minutes discussing how one might respond instructionally to these confusions. Then we share these responses, with the assigned group explaining their instructional moves and their responses. This process enables the whole class to begin to consider the purposefulness of instructional moves. Taking the time to discuss each of these confusions enables us to build a public base of knowledge about the work teachers must do to anticipate and respond to their students' ideas. We repeat the process with the understandings that students in the case show. In the next section, I describe what participants generate during each of these discussions.

What Teachers Learn from Anticipating and Responding to Student Thinking

In this section, I summarize some of the insights that teachers gain by engaging in an analysis of students' responses to the multiplication cluster problems. We typically begin with the confusions encountered by students in the case that teachers identified, including (a) basic place value issues such as what the 7 represents in 72, (b) not knowing which of the subproducts should be added to get the product, (c) mimicking classmates but not really understanding the process, and (d) adding zero when multiplying by 10 without knowing what that means. The participants' first instinct in responding to any of these confusions is for the teacher to step in and demonstrate how the students should solve the cluster problems. To move them away from this instinct and consider how to help students do the intellectual work, I press them to think about what might be causing students' confusions and what kinds of questions the teacher could ask to gain insight into what is difficult for students. This idea helps them consider how to talk to students who are mimicking their classmates or who are not sure which subproducts to add in order to learn what they *do* understand. Teachers suggest asking students to put in their own words what a cluster problem is. Questions like, "What are we trying to figure out here? Tell me, what you do understand?" are considered useful. Another idea teachers generate is to create an easier cluster problem to determine if a simpler problem can clarify the purpose of the task before jumping to harder numbers. Teachers consider asking students to draw an array for the target problem (such as 23×4) and then separate arrays to show each of the suggested subproducts. Then, they suggest that students look to see which of the subproducts can be used to fill up the total array and how. Drawing arrays and looking for subproducts is seen as a potentially powerful way to develop an understanding of what happens when you multiply by 10 and append zero to the answer.

When we discuss students who show confidence with cluster problems, I also press teachers to think about alternative ways of challenging students beyond simply making the numbers larger. Teachers generate ideas, such as asking students to look for regular patterns in ways that cluster problems are broken up and to think about whether those patterns are true for all numbers. This might lead students to think about whether doubling and halving will always work, or whether decomposing a number into tens and ones and multiplying by those subproducts will always work. I ask them to think about the logic students who are confident with cluster problems use when they generate their own. Can they create a cluster, for example, by dividing the problem into just two sub-products? What about three? What are efficient methods for accomplishing this?

Teachers have found this close attention to the range of student thinking to be helpful in generating ideas about how to engage with students' confusions and understandings. Teachers have been particularly interested in expanding their ideas about how to respond to students' ideas beyond merely stepping in and doing the work or re-explaining the problem.

Conclusion

Because cases are situated in real classroom situations, they are likely to create animated discussions. We take seriously the call in the cases literature to think strategically about the purposes case discussions can serve and how to facilitate them to promote learning (Levin, 1995). In this chapter, we have described our efforts to focus discussions in particular ways to bring to the surface *how* teachers use their knowledge of mathematics and student thinking to plan for and to facilitate instructional activities. Not all of the DMI cases are well suited to examine the teacher's role in using children's thinking to guide instruction. When making these adaptations, we looked at whether the text of the case allowed us to create discussion questions that would focus teachers' attention on pedagogical choices and decisions. The cases vary in the extent to which the reader has access to the flow of a classroom lesson or the conversations that occurred between teachers and students. Those features are the raw material we need in order to craft case-based discussions that pay explicit attention to how teachers make decisions as they design tasks and interact with students.

Each of the adaptations we presented in this chapter show a different way to make use of those raw materials. In the first adaptation, we paired two cases that would ordinarily not follow one another in order to help teachers think about how their choice of task can support students' thinking. In the second adaptation, the dialogue between the teachers and a student produced a useful context to focus on teacher questioning. And finally, in the third adaptation, the text of the written case included an array of confusions and understandings that emerged from one lesson so that we could focus on how a teacher might anticipate and respond to that range of thinking. Each of these adaptations highlighted a core task of teaching: designing tasks, eliciting student thinking, anticipating and responding to student strategies. By documenting and describing how teachers with whom we have worked have engaged in specific discussions around these three adaptations, we have shown how our work helps teachers build knowledge about what they do when using students' mathematical thinking to guide instruction.

References

Broudy, H. S. (1990). Case studies—Why and how. *Teachers College Record, 91,* 449-459.

Economopoulos, K., Russell, S. J., & Tierney, C. (1998). *Investigations in number, data, and space.* Menlo Park, CA: Dale Seymour.

Harrington, H. L., & Garrison, J. W. (1992). Cases as shared inquiry: A dialogical model of teacher preparation. *American Educational Research Journal, 29,* 715-735.

Levin, B. (1995). Using the case method in teacher education: The role of discussion and experience in teachers' thinking about cases. *Teaching and Teacher Education, 11,* 63-79.

Lundeberg, M. A., Levin, B. B., & Harrington, H. L. (Eds.). (1999). *Who learns what from cases and how?* Mahwah, NJ: Erlbaum.

Schifter, D., Bastable, V., & Russell, S. J. (1999). *Developing mathematical ideas: Building systems of tens.* Parsippany, NJ: Dale Seymour.

Smith, M. S., Silver, E. A., & Stein, M. K. (2005). *Improving instruction in algebra: Using cases to transform mathematics teaching and learning, Volume 2.* New York: Teachers College Press.

Stein, M. K., Engle, R. A., Hughes, E. K., & Smith, M. S. (in press). Orchestrating productive mathematical discussions: Helping teachers learn to better incorporate student thinking. *Mathematics Thinking and Learning.*

Sykes, G., & Bird, T. (1992). Teacher education and the case idea. In G. Grant (Ed.), *Review of Research in Education* (Vol. 18, pp. 457-521). Washington, D.C.: American Educational Research Association.

---

[1] Much of our collaboration has been in the context of two grants funded by the National Science Foundation: Expanding the Community of Mathematics Learners (ECML, #9819438) and Strategic Organization, Assistance and Resources for Washington Mathematics (SOAR, #0554541). The opinions expressed here are solely the authors and do not reflect the views of the National Science Foundation.

Elham Kazemi is an associate professor of mathematics education at the University of Washington. She studies and designs professional education to help teachers center instruction on children's mathematical thinking.

Anita Lenges is a faculty member in the Masters in Teaching program at The Evergreen State College. Her work centers on teacher preparation that supports racially, culturally, and linguistically diverse students to thrive in mathematically rich environments.

Virginia Stimpson is a research associate at the University of Washington. She supports schools and districts as they enact long-term plans for systemic change in the teaching and learning of mathematics. Most recently she has been involved in researching leadership content knowledge.

Schifter, D. and Bastable, V.
AMTE Monograph 4
Cases in Mathematics Teacher Education: Tools for Developing Knowledge Needed for Teaching
©2008, pp. 35-46

4

Developing Mathematical Ideas: A Program for Teacher Learning[1]

Deborah Schifter
Education Development Center

Virginia Bastable
SummerMath for Teachers at Mt. Holyoke College

The principles that, in our view, underlie effective teaching practices are such abstract directives as: know the mathematics, center your lesson on student thinking, and try to discern a logic in students' mistakes. For most teachers, learning to translate these principles into competent management of the contingent, moment-to-moment character of classroom interaction is a process fraught with both intellectual and emotional challenges. Developing Mathematical Ideas (DMI) is a curriculum for K-8 teacher learning designed in response to this challenge. It uses case discussion to model how close attention to student thinking and a deep grasp of the mathematics at issue can lead teachers to a more coherent, hence more confident, approach to their instruction. DMI consists of seven modules, each focused on a particular mathematical theme studied over eight 3-hour seminar sessions. It integrates case discussion with opportunities for participants to explore relevant mathematics, investigate the mathematical thinking of their students, analyze lessons from innovative curricula, and consider related research from the education literature. This chapter excerpts the discussion of a single case from the second session of a seminar, Reasoning Algebraically About Operations, to demonstrate how teachers can come to recognize that questions about pedagogical strategy should be based on and follow from consideration of the mathematics content and student thinking.

The principles that, in our view, underlie effective teaching practices are such abstract directives as: know the mathematics, center your lesson on student thinking, and try to discern a logic in students' mistakes. For most teachers, learning to translate these principles into competent management of the contingent, moment-to-moment character of classroom interaction is a process fraught with both intellectual and emotional challenges. For example, as teachers begin to enact a practice in which student talk is foregrounded, they often seek reassurance and ask for lists of "good" questions. However, questions out of context are unlikely to be useful.

By studying cases, teachers can begin to learn "good" questions do not exist in isolation from the flow of classroom conversation. Teachers realize that effective instruction depends upon a deep grasp of the mathematical goals of the lesson; bringing students from where they are to the mathematics to be learned requires careful attention to student thinking. Case discussions model how such attention to students' ideas, coupled with the ability to identify significant mathematics, can help teachers develop a feel for questions that open important areas for investigation. In short, though "right" questions cannot be anticipated, teachers can learn to approach their mathematics instruction with a coherent set of priorities suggestive of productive questions.

We recognize that other forms of professional development provide their own learning opportunities. Thus, it is important for teachers to engage in mathematical investigations that allow them to experience the challenge—and satisfaction—of understanding content that has previously eluded them. We want teachers to learn how to analyze lessons in published curricula, and to learn how to inquire into their own students' thinking.

Furthermore, we see the benefits of coordinating such professional development activities. For example, teachers' own mathematical investigations position them better to interpret the mathematics worked on by students in the cases. At the same time, a set of cases allows teachers to see how the mathematics they are learning can arise in different contexts with students of different ages. Each activity enhances the others.

These considerations led to the production of a professional development series called *Developing Mathematical Ideas* (DMI). DMI consists of seven modules[2]; each module focuses on a particular mathematical theme studied over eight 3-hour sessions. Facilitator guides include mathematics activities, focus questions, and assignments for each seminar session.

Although the richness of DMI comes from the coordination of a variety of activities and connections among concepts studied over time, in this chapter we focus on a single case from the DMI series. We consider how the case discussion provides opportunities for teachers simultaneously to work on a mathematical issue, examine student thinking, and analyze how one teacher's pedagogical strategies respond to her students' ideas. Consider the following case from a kindergarten classroom.

A Teaching Episode[3]

The teacher, Lola[4], has set up her students to play Double Compare[5], a card game similar to War: Each card bears a numeral, from 0 to 6, and a picture of that number of objects. Players lay down the top two cards from their piles, and the player with the higher total when the numbers on the cards are combined says "me." For example, when Wei turns over a 2 and a 6, and Marta turns over a 3 and a 4, they count the totals to determine that Wei has 8 and Marta, 7, so Wei says "me." Lola describes classroom events as the game gets underway.

As I watched, a situation came up with several groups: each partner would have one card equal to the other person's and one card that was different. When Martina had 6 and 2 and Karen had 6 and 1, Karen quickly said "You." I asked how she knew and she pointed to the 2 and said, "This is big. Even though these are the same [the 6s], this [the 6 and 2] must be more."

Paul and his partner had a very similar set of hands. Paul put down 6 and 3 and his partner put down 6 and 1. Paul said, "I had 6 and he had 6, and then I had a higher number." I asked what their cards added up to, and both of them counted all the little pictures on their cards to get the totals. As they continued to play they did not count to figure out totals or who had more, but did accurately figure out who got to say "me." On a turn a minute later, Paul had 4 and 3 and his partner had 4 and 5. Paul's comment was, "I have 3 and he has 5." He knew he could basically ignore the two 4s. Another hand I saw was when Karen had 6 and 5 and Martina had 0 and 2. Karen said "Me, because she got two low numbers."

After a while, I realized that it seemed almost no one was EVER adding or counting or figuring out totals. I looked around some more, mostly just collecting data mentally about whether I saw adding, counting, or discussion of totals. I saw just a tiny little bit.

It was time to clean up and meet on the rug. Once we were all settled, we talked about how students knew who got to say "me." We talked for a while about how all the pairs "ignored" cards when each partner had the same one, and only paid attention to the cards that were different. Martina said that 6 and 3 is more than 6 and 1 because the 3 is bigger than the 1. I asked, "What about the sixes?" and she said, "They're the same." Paul added, "They don't matter. You don't have to pay attention to the sixes." I pointed out to them that when I put down the 6 and 1, they said, "That's seven," but when I put down the 6 and 3, no one figured out what it made. "Would 6 and 3 make a higher number than 6 and 1?" I heard 8! 9! 10! They settled on 9 by counting all, and because Danielle said 6 plus 3 is 9. "Is 9 more than 7?" Yes!

These students seem to have made a generalization, that a number plus a big number is more than the same number plus a small number. I put out a few more sets of cards, varying the number that was the same

("Does this only work for 6?" "No.") They said it always works, and Paul reiterated that you don't have to pay attention to the numbers that are the same.

40 Another generalization most of them seemed to be using was something about two small numbers is less than two big numbers. Karen's comment that she got two low numbers expressed this idea. I asked the group about this. I put out 1 and 5 and 0 and 4, which had been a turn in Amanda and Danielle's game. I asked how Amanda 45 knew she had more. She said, "This [5] is bigger than this [4], and this [1] is bigger than this [0]." I asked if it would work with other numbers and everyone said yes. We tried some. They were all saying it worked. They weren't adding and counting. They were "just knowing."

Implicit in the children's actions were two generalizations. For 50 one, the children were close to articulating what it was: "You don't have to pay attention to the sixes." I wonder what it will take for them to have words for their second generalization beyond simply saying they "just knew." Before moving to the case discussion, consider what this case illustrates. In particular, consider how Lola's actions are 55 based in her knowledge of mathematics and her attention to student thinking.

Double Compare was designed to give young children an opportunity to practice counting to find totals. Yet, as they begin to play the game, it seems some of Lola's students are subverting that goal of the exercise: They play many rounds correctly, but without totaling.

However, Lola quickly moves beyond noticing what her students *aren't* doing to figure out what they *are* doing. Paying careful attention to their strategies, she realizes that implicit in their moves is a general principle: If each child has a card of equal value, compare the other two cards; the child with the card of greater value has the greater total. When it is time to put the cards away, Lola brings her students together to discuss this idea.

Lola has a sense of the important role of generalization in the doing of mathematics—and she recognizes that even kindergartners can participate in such a central mathematical activity. As young children learn about numbers and operations, they begin to see regularities in our number system, and to think about what stays the same among things that are changing. The regularity, or generalization, illustrated above—if one number is greater than another, and the same number is added to each, the first total will be greater than the second—is one example. This statement is true for *any* three numbers. For example, since $54 > 36$, $54 + 98 > 36 + 98$. This idea can be expressed in algebraic notation: *For all numbers a, b, and c, if* $a > b$*, then* $a + c > b + c$.

However, it is important not to attribute too much understanding to these kindergartners. The children are playing a card game that involves the numbers

0 through 6. They are to determine who gets to say "me" by comparing the number of pictured objects when they each turn over two cards. We do not know if the children have a conception of numbers greater than 6, whether they think in terms of the operation of addition, or whether the ideas they are discussing apply to contexts outside the card game.

Yet as this episode illustrates, teachers can invite even very young children to talk about the general ideas implicit in their actions. Consider Lola's moves: First, she notices that some of the children are playing the game in a way that indicates they may be acting on a general principle or rule which they notice but have not articulated. Second, she decides to bring this principle to the attention of the whole class during discussion. Third, she asks her students to explain their rule and why it works by suggesting additional examples for them to consider. Finally, she poses the question: Does this rule work for just a few examples or does it apply more generally?

Having paid careful attention to the ideas her students are engaging, Lola situates her students' actions in the context of important mathematics. She presses them to explain themselves as they work in pairs, and then raises questions for whole-group discussion. She not only brings to their attention the mathematical generalization implicit in this activity, but also begins to inculcate the habit of looking for rules and regularities more generally.

A Sample Case Discussion

The following excerpt from a DMI seminar illustrates how case discussion can provide a context for teachers to deepen their mathematical understanding, sharpen their focus on student thinking, and examine the pedagogical moves that follow from these. Participants in the seminar were practicing teachers of first grade to middle school. In this excerpt, the teachers are discussing the previous kindergarten case, "Double Compare," which takes place in the second session of the seminar, *Reasoning Algebraically About Operations*.

Deepening Mathematical Understanding: Expressing Generalizations

One goal of the *Reasoning Algebraically About Operations* seminar is to learn to identify and articulate mathematical generalizations in both natural language and algebraic notation. This, in turn, serves several purposes: 1) teachers will be better prepared to notice when their students broach important mathematical themes, 2) teachers can identify different levels of generality and consider the extent of the generalization their students may be formulating, 3) teachers can learn the conventions of algebraic notation, 4) teachers can vest algebraic notation with meaning, because they are using the notation to formulate ideas already articulated, and 5) teachers can learn to appreciate both the precision and conciseness of the algebraic notation.

To illustrate how teachers take initial steps toward these goals, consider the following segment of the case discussion. The facilitator begins by asking for the generalizations Lola's students were developing.

Evelyn speaks up first, "This is what I wrote for Paul. He is talking about the cards he and his partner have, 4 and 3 versus 4 and 5. I wrote this for what Paul did: If two cards are the same, just ignore them and compare the other two numbers."

Lucy agrees, "That was on my list of generalizations, too."

In order to challenge participants to produce a more precise statement, the facilitator suggests pairs of cards that satisfy Evelyn's condition, but which she had not intended to include: "How about this situation, 2 and 3 versus 5 and 5? Would that apply? It matches the words Evelyn used, 'If two cards are the same, just ignore them.'"

Evelyn refines her statement, "You have to say, if each of us has a card that is the same, then look at the cards we have that are different."

Beverly refers back to Lola's language in the case and says, "I think there is another way to say it. It seems like the same idea but also a bit different. A number plus a big number is more than the same number plus a smaller number. It might be easier with some letters."

The facilitator proposes that the group consider another set of cards: "If one person has 5 and 3, and the other person has 5 and 2, which is the big number?"

Beverly sees her point, "Oh, yeah. In that case, 3 is the big number, even though it's smaller than 5."

Evelyn adds, "I tried symbols, too. It was odd because it was almost easier with symbols." She comes to the board and writes, "$n + a > n + b$ when $a > b$."

Because Evelyn has just introduced an idea related to one of the seminar's central goals, the facilitator turns to the group and asks, "What made it seem easier?"

Naomi: "You can tell what 'bigger number' refers to, and you can say 'n' for the same number without having to write it all out."

Then Nancy brings the group's attention to another part of the case, "I think there is another generalization here. How about this: Two small numbers added together is less than two big numbers added together."

The facilitator decides not to pursue a more precise statement in English and asks instead if anyone has written a version of Nancy's words in algebraic notation.

Linda comes to the board and writes, "*If* $a < b$ and $c < d$, then $a + c < b + d$." She then turns around and says, "But I am wondering, in the card game, the numbers are from 0 to 6. Do we need to say that too?"

Linda has raised a significant issue: To what domain does the generalization apply? And in the context of this seminar, the question has two forms. 1) What is the domain the students are considering? 2) In what domain is the generalization true? To conclude this portion of the discussion, the facilitator says, "Linda has made an important point. If we assume these statements are based on the context of the game in the case, then the variables are the whole numbers from 0 to 6, but are the statements true more generally? What are the numbers we can choose for a, b, c and d and still have the statement be true? When we write statements in symbols, we need to consider the domain. We will look into this more as the seminar goes on."

In the first part of the case discussion, participants share both symbolic and natural-language versions of the generalizations they uncover in the thinking of Lola's kindergartners. They use the notation so their statements accurately reflect their own ideas. In addition, participants note that symbolic statements are sometimes easier to write and understand than the natural-language versions. Finally, they begin a conversation about the domain the variables satisfy, something that will be examined in more detail as the seminar continues.

Focusing on Student Thinking: Are Students Thinking About Operations?

The seminar discussion illustrates that as participants engage the mathematics, the teachers are also paying careful attention to student thinking, studying the kindergartners' words in order to uncover the general principles underlying their strategies. When Linda asks her question about the domain, the facilitator uses the opportunity to highlight the boundary between the students' ideas (they might be thinking only in terms of the numbers 0, 1, 2, 3, 4, 5, and 6) and the general mathematical principles the teachers are sorting out for themselves (the generalization is true for all real numbers).

As discussion continues, the teachers extend the generalization in another direction—now to consider other operations—and then return to the kindergartners to consider more specifically the mathematics they are exploring.

Kathleen begins this part of the discussion by comparing Lola's case to the work of her third graders: "You also need to think about what operation this works for. My class is working on arrays and there's a game where they compare array cards. I hear kids saying something very similar, 'I know if this is 6×5 and 6×8, I can just ignore the 6s and compare the other numbers.' When I asked them to explain, they say, 'You have 8 rows of 6 chairs or 5 rows of 6 chairs. All you need to do is see how many rows you have and then you are all set.' So then the question is, in which operations can you do this? When can you ignore one of the numbers and just compare the others? How do the kids get to know that?"

Kathleen raises an important issue. She points out that her third graders are working on an analogous problem. Adapting Evelyn's formulation in algebraic notation, one might represent the generalization implicit in the third graders' explanation as, "For all counting numbers n, a, and b, $n \times a > n \times b$ when $a > b$." (Note that the context of the third graders' thinking is arrays, which necessarily involve counting numbers.)

But Kathleen has asked, when students make such generalizations, are they thinking in terms of a specific operation—about joining two sets (in the case of the kindergartners), or about multiplying (in the case of the third graders)? Or are the students thinking that under any operation, if one number of each pair is the same, you can compare the remaining numbers?

Later in the seminar session, participants will discuss another case in which fourth graders investigated the same question with regard to subtraction[6]. Comparing $145 - 100$ and $145 - 98$, it seems to the students counterintuitive that $145 - 98$ gives the *larger* result, when 98 is *less than* 100.

Knowing that the teachers will have an opportunity to discuss the subtraction case, the facilitator underscores Kathleen's point while suggesting the group will return to this issue: "The questions Kathleen is asking are important ones for us to keep in mind. Once a generalization is noticed in one context or with one operation, what happens if the operation changes? Will it still be true? Why or why not? As students work on such questions, they learn more about the operations. We can, too. Let's be sure to mark these questions so we can come back to them."

Linda brings the conversation back to the kindergartners. In contrast to Kathleen's third graders, Lola's students can't consider other operations yet: "Since this is a kindergarten class, they only have counting and maybe adding."

Nancy objects, "But they're just looking at the cards. It doesn't seem as much like addition."

Evelyn disagrees, "It is because they know they are putting amounts together they are saying you can ignore the part that's the same. If they weren't adding them, then what are they doing? They know they don't have to actually find the total to know which combined amount is greater. The whole context is addition."

But now Beverly objects, "I'm not sure if the children are really thinking about addition. In lines 45-46, Amanda just seems to be matching and comparing. Amanda says something like, 'This is bigger than this and this is bigger than this.' It feels like she is just looking at the cards and not doing any adding."

Lucy expresses bewilderment, "In the case, Lola was saying they weren't adding or counting. They were 'just knowing.' I keep going back and forth myself. What are they doing? Counting? Comparing? Adding?"

Kathleen responds, "I think Lola's point was that they didn't have to find the total in order to compare. So in a sense, they weren't counting or adding. But they did seem to understand that they were asked to compare the results of a combining situation."

The question the participants are debating isn't resolved. Instead the facilitator asks, "Why is this a useful question to consider? How does it help us understand what is important for students to learn?"

There is a pause in the discussion. In fact, it isn't necessary for participants to answer the question at this time. But it is an important question to consider: Why does it matter whether the kindergartners are thinking in terms of the operation itself or just thinking about cards. Or, more generally, why should teachers work so hard to understand the thinking of these students?

Naomi contrasts the discussion with her own practice, "It is really odd for me to think in so much detail. With my middle schoolers, I am not used to noticing the kinds of differences Lucy is talking about."

Rather than answer Naomi's implied question, the facilitator encourages Naomi to take on this task of listening: "Listening closely to student ideas is something to explore throughout this seminar. For each session, the cases offer you opportunities to think about students' ideas in detail. But you are also asked to do assignments to record the mathematical conversations that take place in your own classroom. This is an opportunity for you to examine the ideas of *your* students."

Investigating Teacher Moves: Guided by Mathematical Understanding and Attention to Students' Ideas

The normative mode of professional sharing in current school culture involves teachers showing one another "activities that work," and analysis rarely moves beyond superficial and evaluative reactions: "I like what she did when . . ."; "Instead, she should have . . . ." In the initial meetings of a DMI seminar, the facilitator must help participants learn to examine the cases for what they reveal about mathematics and children's mathematical thinking: What is the mathematics at play in this lesson? Based on the evidence (what a child says, does, or writes), what might we hypothesize about what the student understands/does not understand? What is the mathematical idea he or she seems to be using? Questions about pedagogical strategy are *based on* and *follow from* consideration of the mathematics content and student understanding.

As we shall see in the final excerpt from the discussion of the "Double Compare" case, the teachers in this seminar have moved beyond judging Lola's pedagogical actions to consider how her moves grow out of careful attention to student thinking and, informed by her knowledge of the mathematics content, her goals for the lesson.

Nancy opens this part of the discussion, "The teacher let the children play the game their way. She doesn't stop them and make them count and add even if that is what she expects."

Evelyn continues, "From the beginning of the case, Lola looks around the room and sees that it isn't just a few students but most of the class is working this way. It feels like that is a moment when she is assessing what is going on and deciding what to do about it. Then she calls them all together to discuss it, doesn't she?"

Beverly adds, "She doesn't just observe what they are doing, but when she pulls them together she makes that the topic of conversation. She comes up with examples for them to look at that bring the idea out. She really thinks fast."

Wanting Beverly to be more precise, the facilitator asks, "What is the idea you are talking about, Beverly?"

"That sometimes you don't need to add to determine who has the most."

Evelyn specifies, "Around lines 18 to 21, Lola seems to say the point of the game is about finding totals—counting and adding." She continues, "By the time she gets to the whole group discussion she has changed her objective. Now she wants them to be clear about how they could get the answer when they weren't adding."

Nancy elaborates Evelyn's point, "Lola notices they are not doing what she thought they would, but she also sees the value, the mathematical importance, of what they *are* doing. I mean she could have just thought, 'Well. This game isn't going very well. Let's put it away.' But she doesn't. She is sensing there is something else here."

As discussion continues, participants now consider how Lola conducts the whole-group discussion. But at this point, we have seen enough to form a good idea of what aspects of the lesson receive the teachers' attention. They see that, from the start of the Double Compare activity, Lola carefully observes her students as they play the game; that most students don't perform the task as expected; that she is quick to see the mathematical value in the children's actions; and that subsequently Lola focuses whole-group discussion on the mathematical principles underlying her students' strategies.

Facilitating a Case Discussion

The principles that underlie effective professional development for teachers are analogous to those for teaching K-12 students: Effective instruction depends upon a deep grasp of the goals for the course as well as the individual lesson; and bringing students (who may be teachers) from where they are to what is to be learned requires careful attention to their thinking.

For example, consider the portion of the discussion in which teachers worked to state the generalization implicit in the kindergartners' activity. The facilitator articulated a set of goals for the teachers to achieve over the course of the seminar, and understood how this particular case discussion would support it. Ideally, each of her actions is intended to serve those goals.

- As teachers proposed ways to state the generalization, the facilitator pressed for greater precision, providing examples that satisfied the conditions of their statements but were not illustrations of the generalization they were trying to articulate.
- When teachers broached an important issue, the facilitator probed to bring them deeper into the idea: When Evelyn said that it is easier to state the generalization with symbols, the facilitator asked, "What made it seem easier?"
- When Linda pointed out that the kindergartners were working only with the numbers 0 to 6, the facilitator recognized that she was raising a significant mathematical point about the domain under consideration. The facilitator explained the importance of Linda's observation, introduced terminology, and informed the teachers that this issue would be explored more deeply later in the seminar.

To summarize, although teachers' learning may begin by reading the case, it is through carefully facilitated discussion that they engage with the ideas whose investigation the instructor has set as the goal for the lesson.

Conclusion

The chapters of this book illustrate a variety of approaches to case discussion. The use of cases in DMI is shaped by our view of effective mathematics teaching, what our experience has led us to believe are the knowledge and skills required of teachers to enact such a practice, and a sense of how teachers might learn these.

Specifically, we start with a view of instruction that foregrounds student ideas along with clear goals for student learning. The art of teaching involves helping students move from where they are into the content to be learned. Such a practice depends on teachers having a deep understanding of mathematics content and the ability to situate students' ideas in that content.

For most teachers, learning to translate the abstract principles that underlie effective teaching practices into competent management of the contingent moment-to-moment character of classroom interactions is a challenging process both intellectually and emotionally. DMI's case methods respond to this challenge by modeling how careful attention to student thinking and a grasp of the mathematics at issue can help teachers develop a coherent approach to mathematics instruction.

DMI offers other forms of professional development activity to supplement the cases. Thus, each module integrates case discussion with opportunities for teachers to learn the relevant mathematics for themselves, analyze lessons from innovative curricula, study the mathematical thinking of their own students, and read about related research in the mathematics education literature. Modules are designed for eight 3-hour sessions, providing for sustained investigation of a major mathematical strand and illustrating central conceptual issues of the K-8 curriculum. Though this chapter describes and analyzes a discussion based on a single case, it is important to note that the context for this discussion is the second session of an eight-session seminar.

References

Russell, S. J., Economopoulos, K., Wittenberg, L., et al. (2008) *Investigations in Number, Data, and Space®*, Second Edition. Glenview, IL: Pearson Scott Foresman.

Schifter, D., Bastable, V., and Russell, S. J. (with S. Monk). (2008). *Reasoning algebraically about operations, Casebook.* Parsippany, NJ: Dale Seymour Publications, Pearson Learning Group.

---

[1]This work was supported by the National Science Foundation under Grant No. ESI-0242609. Any opinions, findings, conclusions, or recommendations expressed in this chapter are those of the authors and do not necessarily reflect the views of the National Science Foundation.

[2]*Developing Mathematical Ideas* is a staff development program for Grades K-8. The following *Developing Mathematical Ideas* titles are products of Pearson Education, Inc. or its affiliate, publishing as Dale Seymour Publications (Pearson, One Lake Street, Upper Saddle River, NJ 07458).

- *Building a System of Tens*
- *Making Meaning for Operations*
- *Examining Features of Shape*
- *Measuring Space in One, Two, and Three Dimensions*
- *Working with Data*
- *Reasoning Algebraically About Operations*
- *Patterns, Functions, and Change*

To find out more about the program or to order, call 1-800-321-3106 or visit www.pearsonschool.com.

[3]The episode is excerpted from "Double Compare," which appears in the *Reasoning Algebraically About Operations, Casebook*, Schifter, et al., 2008, pp. 31-33.

[4]Pseudonyms are used for teacher and students.

[5]The game, Double Compare, is taken from "How Many Do You Have?: Addition, Subtraction, and the Number System," a kindergarten unit of *Investigations in Number, Data, and Space* (Russell, et al., 2008).

[6]"Is it two more or two less?" in *Reasoning Algebraically About Operations*, Schifter, et al., 2008, pp. 45-47.

Deborah Schifter is a principal senior scientist at the Education Development Center, Newton, MA. She has worked as an applied mathematician; has taught elementary, secondary, and college level mathematics; and, since 1985, has been a mathematics teacher educator. She has a B.A. in liberal arts from Saint John's College, Annapolis, an M.A. in applied mathematics from the University of Maryland, and a M.S. and Ph.D. in psychology from the University of Massachusetts. Dr. Schifter co-authored with Catherine Twomey Fosnot *Reconstructing Mathematics Education: Stories of Teachers Meeting the Challenge of Reform* and edited a two-volume anthology of teachers' writing, *What's Happening in Math Class?* She is an author of the second edition of *Investigations in Number, Data, and Space*, in particular having worked on the algebra strand. In collaboration with Virginia Bastable and Susan Jo Russell, she has produced the professional development series, *Developing Mathematical Ideas*.

Dr. Virginia Bastable has been the Director of the SummerMath for Teachers program at Mount Holyoke College since 1993 after working as a secondary mathematics teacher for more than twenty years. She has a B. S. in mathematics with a minor in education from the University of Massachusetts, a M. S. in Secondary School Mathematics Teaching from Worcester Polytechnic Institute, and an Ed. D. in mathematics education from the University of Massachusetts.

Since 1992, Bastable has collaborated with Deborah Schifter of EDC, Inc. and Susan Jo Russell of TERC and many reflective teachers to produce the *Developing Mathematical Ideas* (DMI) professional development curriculum. Bastable and Schifter also lead the DMI Leadership Institutes designed to provide support for teacher-leaders, math specialists, math coaches and others who work in school systems to help educators reconceive the way mathematics instruction takes place. She is an author of the second edition of *Investigations in Number, Data, and Space* with a focus on algebraic thinking in the number units of grades 3-5 and in the Patterns and Functions units.

Henningsen, M. A.
AMTE Monograph 4
Cases in Mathematics Teacher Education: Tools for Developing Knowledge Needed for Teaching
©2008, pp. 47-56

5

Getting to know Catherine and David: Using a Narrative Classroom Case to Promote Inquiry and Reflection on Mathematics, Teaching, and Learning

Marjorie A. Henningsen
American University of Beirut

This chapter describes the use of a narrative case of two contrasting lessons that deal with algebraic thinking in the context of a university elementary mathematics methods course. Data are presented to shed light on what preservice teachers might learn from reading the case, discussing it, and engaging in a variety of mathematical and analytical tasks related to the case. The value of the case lies in the fact that it can be used to explore mathematical ideas, pedagogical issues, and the benefits of teachers reflecting on their own practice.

Cases have been used for professional education in the fields of law, business, and medicine for nearly 150 years (Merseth, 1996). However, their use in the education of teachers emerged within the past few decades and gained increasing prominence since the late 1980s. Cases used in teacher education come in a variety of forms, including video and audio, CD-Rom, transcripts, narrative re-telling, and artifacts or records of practice. Cases may represent large or small chunks of teaching practices, they may be authored from a variety of perspectives, and they may or may not be designed to highlight specific aspects of educational practice or to be exemplars of more general ideas about teaching and learning. Cases can preserve the complexity of teaching, that is, they can illuminate an instance of authentic practice as it unfolds in a context (Henningsen, Stein, Smith & Silver, 2000; Smith, 2001). However, cases and other records of authentic practice are not self-enacting; rather, they form the basis from which appropriate learning tasks can be designed (Ball & Cohen, 1999; Smith, 2001).

The present chapter focuses on the use of one particular narrative case in the context of an elementary mathematics methods course in which cases of various types were used to enhance teachers' mathematical understanding and explore important pedagogical ideas. The narrative cases used in the course are lengthy written representations of entire middle school mathematics lessons and the context in which they took place. Each narrative case provides extensive information about the school context, teacher experience, broad and specific curricular and teacher goals, a full narrative summary of the lesson, and examples of student work from the lesson. At the same time, the cases are specifically designed to bring to the fore and provoke discussion about particular mathematical ideas and representations, and also to serve as exemplars of certain empirical pedagogical patterns related to how students in the case are supported to think, reason, and communicate at a high level. Two important questions to consider are (a) What, if anything, do teachers learn from the case experience? and (b) How can we find evidence of their learning? Within the context of a methods course it is not easy to collect evidence of what teachers learn specifically from the use of particular cases. In this case enactment, however, a deliberate attempt was made to collect evidence through teachers' work on related mathematical tasks before and after the case discussion, individual written analyses of the case, artifacts from small and whole class discussions, and individual written self-assessment of their own learning.

Exploring these Questions with Real Data

This study draws on an enactment of "Pattern Trains: The Case of Catherine Evans and David Young"[1] with preservice elementary teachers enrolled in an elementary mathematics methods course in a university. The specific data discussed in this paper are drawn primarily from the work of 21 students taking the methods course. The course consisted of 15 meetings of 2.5 hours each. Materials related to Catherine and David were used in Session 5 (1.5 hours on the hexagon task) and Session 12 (2 hours Case Discussion and Self Assessment after students had read the case) in addition to a review of the task and written homework assignment in Session 11 in preparation for Session 12. Other cases in both narrative (three others) and video form (total of four) were used during the course to bring to life the mathematical and pedagogical ideas covered during the course. The Catherine and David case was primarily used to enhance our discussion of teaching algebraic thinking in elementary school.

The Case Materials and Learning Tasks

The case materials consisted of two parts. First there was a mathematical activity in which learners explored the growing visual pattern shown in Figure 1. With each successive hexagon train, one new hexagon is attached to the previous train. So the first train consists of one hexagon, the second train contains two adjacent hexagons, the third train contains three adjacent hexagons, and so on. Teachers were asked to determine the perimeter for the tenth hexagon train without constructing it, and then write a description that could be used to compute the perimeter of any train in the pattern, using one edge length of another pattern block piece as the unit of measure (any piece could be used except the long side of the trapezoid piece).

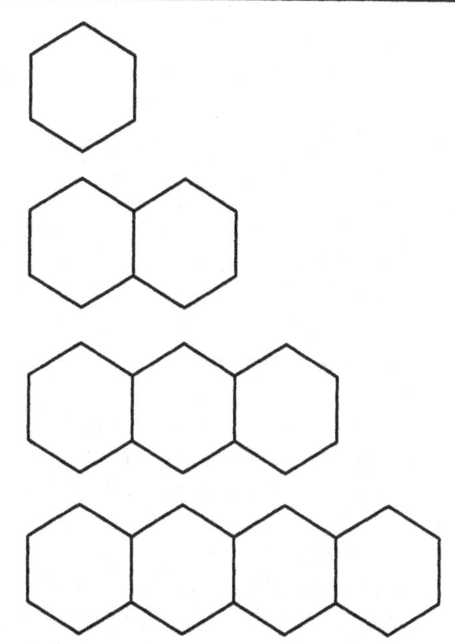

Figure 1. First four figures from the hexagon pattern train task

The second part of the materials used in preparation for and during session 12 consisted of the narrative description of what happened in two different lessons using the same task(s). The narrative itself was divided into three main sections: (1) general background information about the school context in which the two teachers were working, (2) detailed account of a lesson taught by Catherine followed by excerpts from Catherine's journal at different points in time following the lesson (over time she becomes less satisfied looking back on the lesson); and (3) detailed account of what happened in David's lesson when he engaged his students in the same tasks as Catherine. In this case, David was more successful at engaging his students in high-level thinking, reasoning, and communication. Taken as a whole, the case illuminates ways in which a challenging task can be enacted in classrooms, affording different engagement and learning opportunities for students, and how reflecting on one's own teaching can lead to important insights that might affect future practice.
The richness of this type of case supports a diverse array of interesting potential learning tasks for teachers. The 21 teachers were engaged in four main learning tasks surrounding the case materials.

Hexagon pattern task. The hexagon pattern task was completed prior to reading the case in order for the reader to experience and explore the key mathematical ideas as a learner prior to reading how the enactment of the task unfolded in a real classroom. Teachers worked on the hexagon task several weeks prior to encountering the case under the auspices of learning about different strategies for promoting student-student communication in the classroom. They solved the task individually and then interviewed one another to understand their partner's thinking and reasoning on the task. They worked in pairs on the task for about 30 minutes, followed by an hour-long whole class discussion of several solutions and justifications in which teachers reported on their partner's thinking rather than their own. Thus, the task was used as a mathematical learning task, but it was also used as a context for exploring what can be learned by trying to understand another's thinking and reasoning processes. In another session prior to session 12, teachers spent part of the time working on a set of pattern tasks involving both repeating and growing patterns (visual and numerical) most of which can be found in chapter 15 of Van de Walle (2007) which focuses on developing algebraic thinking. However, most of our time in class was spent discussing the numerical pattern sequences because of an interaction between time constraints and teachers' questions during the session.

Read case & write individual reflection. In Session 11, as we began to discuss methods and tasks related to algebraic reasoning in elementary and middle school, we revisited the hexagon pattern task. Solutions were reviewed during the week prior to when preservice teachers were assigned to read the case. Teachers read the Case of Catherine and David and took notes in order to respond individually to two written reflection prompts: (a) To what extent do you agree with Catherine's own assessment of her lesson? and (b) Compare and contrast the quality of mathematical communication in the two lessons. The prompt in part (a) referred to a section in the case in which Catherine's own assessment of the lesson is shown at three different points in time: immediately following the lesson, several weeks later, and months later at the end of the school year. Catherine was quite pleased immediately following the lesson; but with successive viewings over time, she became more critical of her own practices during the lesson and expressed a profound change in her beliefs about what it means for students to be successful in mathematics. The reflection assignments were intended to be completed prior to the case discussion in class; however, eight of the twenty-one teachers actually completed their written reflections after our class discussion. This serendipitous event was used to look for evidence of possible impact of the class discussion on reflections about the case.

Case discussion. In Session 12 teachers were first asked to discuss in small groups the similarities and differences between the two lessons in the case. In order to facilitate the discussion, groups were provided with a Venn diagram recording sheet showing two large intersecting circles, one labeled "Catherine" and one labeled "David." They worked in small groups for about 20 minutes followed by a lengthy discussion of the similarities and differences with the professor recording group ideas on the board. To the extent possible, teachers' ideas were eventually grouped into related categories to encourage them to move from the particularities of the case to more general ideas about teaching and learning that might be embedded in the case. Details about this discussion are described in the following results section.

Self-assessment. Following class discussion, teachers completed a two-part self assessment. In the first part, they were asked to write about what they think they learned from all the activities related to the Catherine and David case (they were asked to be as specific as possible) and what specific tasks related to the case contributed most to their learning. The second part consisted of a growing visual pattern task in which they were asked to find and justify at least two different ways of finding a formula to describe the pattern. The task involved arrays of dots beginning with 2 × 3, then 3 × 4, then 4 × 5, then 5 × 6, and so on. The teachers were asked to write a verbal description of the 10th figure (not shown), to find an expression or formula for the total number of dots in any figure, and to show how they derived their expression. They were asked to find at least two different expressions (e.g., $(n+1)(n+2)$ or $(n+1)^2 + (n+1)$ or $2(n+1) + n(n+1)$, and so on). Teachers had an additional open-ended self-assessment opportunity at the end of the course to describe experiences that had contributed to their learning throughout the course.

Relating Course Goals and Case Learning Tasks

To begin to shed light on what teachers might have learned from the case experience, it is important to articulate how the use of the case fit into the larger course context and why the use of the case was appropriate in the first place. Figure 2 shows how each of the case learning tasks corresponded with the relevant general and specific course goals.

Relevant Overall Course Goals	*Catherine and David - Related tasks*
To analyze and reflect on teaching	Read Case Contrast Catherine and David (written) Venn diagram activity and discussion
To understand the importance of studying patterns (algebra as a study of patterns and functions)	Read and Discuss Case Pair work on hexagon task Discussion of task solutions
To understand the value of collaboration	All of the above
Specific Goals	
To recognize problems in Catherine's teaching and appreciate her ability to reflect on them	Read case React to Catherine's reflections Contrast Catherine and David (written & discussion)
To analyze reform teaching critically	React to Catherine's reflections Contrast Catherine and David (written & discussion)
To reflect on features of real lessons involving pattern tasks	Contrast Catherine and David (written & discussion)
To further develop ability to explore, analyze, and generalize growing visual patterns and to justify formulas visually if possible	Work on hexagon task Discussion of task solutions Self-assessment pattern task

Figure 2. Catherine and David related tasks corresponding to general and specific course goals

Results of the Case Enactment

In my view, the case enactment encompasses not only the reading of the case narrative, but also engagement with all the related tasks. Thus, the presentation and discussion of the data is organized according to the major learning tasks with which teachers were engaged to tease out what preservice teachers might have learned from this enactment of the case. This descriptive analysis begins to shed light on the potential for teacher learning that case experiences may provide.

Hexagon pattern task and self-assessment dot pattern task. When teachers first worked on the hexagon pattern task, nearly all teachers were able to calculate the perimeters of the given trains and the 10th train successfully. Four pairs of teachers (38%) were able to articulate a verbal recursive description of how the train perimeters were growing. Only one pair of teachers (9%) was able to build a formula to describe the perimeter of any train (and they were only able to do so after the facilitator gave them a hint about relating the figure number to the perimeters), but their formula was based on numerical information and they were not able to relate it to the figures visually.

Performance was notably different on the self-assessment dot pattern task described previously. The task was given to teachers seven weeks later and just following the completion of all other work on the Catherine and David case. On the dot pattern task, all 21 teachers were able to write a verbal description of the 10th figure and 67% wrote a general verbal description of how the pattern was growing. For example, one teacher wrote, "You grow it with another column and another row each time so that each dimension increases by one so it's always $+1\ x+2$." Another teacher wrote that "the horizontal and vertical of the array go up by one each time with one dimension being one more than the figure number and the other dimension being two more."

Seventy-six percent of the teachers were able to find at least one appropriate expression (nearly all found $[n+1][n+2]$), but only two were able to generate more than one expression; this still represented a marked contrast to the work on the hexagon task seven weeks earlier when only one pair of teachers could find any algebraic expression to describe the pattern growth. Thus there is evidence that teachers learned something in the seven weeks between their first work with the hexagon pattern task and their completion of the case activities. Although it is not possible to attribute this learning directly to their work on the case activities, it is likely there is a connection; except for the initial hexagon pattern task, nearly all our work with such tasks in the course was done in close relation to the discussion of the Catherine and David case. Many of the preservice teachers self-reported that their work with the Catherine and David case contributed to an increased understanding of pattern tasks like the two described here.

Written individual reflection after reading the case. With respect to the first prompt (reflection on Catherine's assessment of her teaching), the majority of teachers (75%) agreed with Catherine, while those remaining had mixed feelings about Catherine's self-assessment. In those instances, the teachers usually felt that Catherine was too hard on herself. This provided the discussion facilitator with an opportunity to push students to provide evidence to support their disagreement with Catherine, thus leading to a deeper discussion of the case. For example, some teachers had the impression that student communication was high in Catherine's class. We were able to delve deeply into specific passages, including reading aloud and role-playing the interchanges between Catherine and her students. Such in-depth discussion influenced many teachers to change their minds.

In their written reflections, teachers addressed a wide array of ideas related to Catherine's teaching they felt were important to discuss from their own perspective. Their written reflections provide evidence that the case experience may have prompted them to think carefully about certain aspects of the case. The following ideas were addressed substantially by at least 25% of the teachers:

- Catherine was too leading/directive (81%).
- She overly focused on correct answers/discouraged multiple solutions (67%).
- She asked only yes/no questions (43%).
- She overly focused on students feeling good about themselves (38%).
- Choral responding by students can mask real understanding (38%).
- There was a lack of time for exploration (29%).
- Self-reflection can lead to better teaching (43%).

There was not much difference between the ideas addressed by teachers who wrote their reflections prior to the discussion and those who completed the assignment later, except that a higher proportion of teachers who wrote after the discussion focused on the value of teacher self-reflection and also on Catherine's excessive focus on students feeling good about themselves. These two ideas were salient in the whole class discussion.

With respect to the second prompt (to contrast communication in the two lessons), at least 25% of the teachers chose to contrast the two lessons according to the following ideas: student to student communication (76%), students active vs. passive (62%), exploration of multiple solutions (52%), pressure for student explanation (33%), and use of both right and wrong answers as learning opportunities (29%). Teachers who wrote after the discussion were much more likely to discuss ways in which the students were active or passive, particularly the advantages and disadvantages of choral responding. This issue occupied a large chunk of the case discussion. Also, 100% of the post-discussion writers focused on the lack of student-to-student communication in Catherine's class as compared with David's. This can be traced directly to a vivid event during the case discussion in which the facilitator challenged a teacher assertion by having the class act out one of the transcripted portions of the case in order to illustrate how much Catherine was dominating discussion in the lesson.

In-class case discussion. The ideas generated by the teachers during the case discussion also provide evidence of what they might have learned from the case experience or at least the issues about which the case prompted them to think. They clearly recognized that both Catherine and David had similar goals to engage students in high level thinking and reasoning and that both teachers set up tasks for students that held potential for this to happen. They identified a variety of ways in which the enactment of the task differed in the two lessons, focusing primarily on teacher-centered vs. student-centered instruction, allowing for open exploration vs. focusing on correct answers and procedures, listening to and building on student thinking, questioning, exploring multiple representations and solutions, and the role of collegial support and teacher self-reflection in improving teaching. In the discussion the teachers also felt it was important to point out that both teachers wanted students to feel successful and build self-esteem. However, a long discussion ensued about what it means for students to be successful in mathematics from Catherine's perspective(s), David's perspective, and the preservice teachers' own perspectives.

Self-Assessment (What they think they learned). Although the teachers addressed a variety of issues in their self-assessments, only the most commonly discussed ideas will be mentioned here. Seventy-one percent of the teachers asserted that the work on the hexagon pattern task helped them see the importance of exploring multiple solutions and pathways for solving problems. A few also elaborated to say they were learning "how to talk about mathematics a lot more" and the importance of struggling and persevering through a problem. Nearly all (95%) felt that reading the case and discussing it were the most valuable activities. About 50% specifically mentioned that the case made them think about what it means for their students to be successful in math, and that they need to rethink this issue. Sixty-seven percent talked about the relative differences in what teachers can learn about student thinking from asking more closed vs. more open questions. Finally, 43% claimed that the case discussion helped them see the importance of reflecting on their own teaching, alone and with colleagues.

A few weeks after this self assessment, preservice teachers were given an open-ended task to identify a few significant ideas about teaching and learning mathematics that they learned from the course, to present evidence of their own learning, and to talk about what course activities contributed to their learning. Nearly every person made an explicit statement that they learned how challenging teaching is and they gained a better understanding of everything a teacher has to think about when trying to help students learn mathematics. About 80% attributed most of their learning to our work with cases of real teaching. A surprising 67% referred specifically to the Case of Catherine and David as a major source of learning. Some notable quotations about Catherine and David included the following:

- that case helped me put all the pieces together and I tried to relate my own teaching to Catherine and David's;
- critically reflecting on Catherine changed my approach with my students, it made me question how I can know whether my students are really understanding;

- C &D gave me an excellent contrast of good and bad questioning, like a benchmark;
- David gave me an example of what it means to be a 'teacher as facilitator';
- I could see how students can really learn new things by communicating with each other;
- I learned how to instigate a conversation among my students;
- I learned that students can feel successful in math without everything having to be easy; and
- the things that happened to Catherine…I've had them (or worse) happen to me and it helped to talk about it and see what David did and come up with ways of changing my approach.

Concluding Remarks

Clearly the case experience reported here was significant for the majority of teachers in the class. The evidence suggests that these teachers primarily learned from the experience how to be more successful with visual pattern tasks, strategies for promoting student communication and for eliciting student thinking, the importance of defining what it means for students to be successful, and how one's definition of success might connect with the way one teaches. The Case of Catherine and David may have been more salient for teachers than other cases (and possibly even more effective) primarily because it features two contrasting lessons that could be thought of as two cases within the larger story. In fact the only other case mentioned by name in their reflections was another dual case that had been used to enhance our discussion of teaching about multiplying fractions. The dual case narrative format enabled me to engage the teachers with rich analytical comparison tasks about the teaching in the two cases, enabling the teachers to abstract larger ideas above and beyond the specific details of any one case alone. Indeed the long-term potential power of using authentic classroom cases resides in looking across many cases to abstract and generalize to larger ideas about teaching and learning than any single particular case can afford.

I suspect that Catherine and David may have seemed more significant for the teachers because they had found the mathematics to be challenging before we delved into the case. The majority experienced personal growth as learners with respect to dealing with growing pattern tasks, something they had never really understood how to handle before. From my perspective, the case addresses algebraic thinking in an accessible way but allows for meaningful exploration of two well-intentioned, yet contrasting approaches to engaging students in mathematical communication and how student contributions to the discourse look different in the two cases. The case of Catherine and David is also powerful in that it addresses one of my major course goals: developing a reflective stance toward teaching. The small section containing Catherine's reflection on the same lesson at three different time points illustrates the impact self-reflection can have on beliefs and teaching practice in a way that is more elegant, respectful, and succinct than any other case materials I have encountered. Preservice teachers can relate to Catherine's learning trajectory about the issue of what it means to be successful because many of them recognize Catherine's initial beliefs as similar to their own (and often the same as what they see in their mentor teachers during practicum). Thus, the Case of Catherine and David allows me to address simultaneously several layers of important understanding about the processes of teaching and learning mathematics with the layers remaining connected and complex, the way they are in real life. I have since used this case many times in the context of methods courses and the reaction from preservice teachers is nearly always the same — they admire David's ability to elicit thinking from students but they always wish they could meet Catherine in person and learn more from her experiences and reflections on her own teaching.

References

Ball, D. L., & Cohen, D. K. (1999). Developing practice, developing practitioners: Toward a practice-based theory of professional education. In L. Darling-Hammond and G. Sykes (Eds.), *Teaching as the learning profession: Handbook of policy and practice*, (pp. 3-32). San Francisco: Jossey-Bass.

Henningsen, M., Stein, M. K., Smith, M. S., & Silver, E. A. (2000). The use of cases in mathematics teacher education: A summary of the COMET invitational conference. Irving, TX: Exxon Educational Foundation.

Merseth, K. K. (1996). Cases and case methods in teacher education. In J. Sikula (Ed.), *Handbook of research on teacher education, 2nd edition,* (pp. 722-744). New York: Macmillan.

Shulman, J. H. (1992). *Case methods in teacher education.* New York: Teachers College Press.

Smith, M. S. (2001). *Practice-based professional development for teachers of mathematics.* Reston, VA: National Council of Teachers of Mathematics.

Smith, M. S., Silver, E. A., Stein, M. K., Henningsen, M. A., Boston, M., & Hughes, E. K. (2005). *Improving instruction in geometry and measurement: Using cases to transform mathematics teaching and learning, Volume 2.* New York: Teachers College Press.

Van de Walle, J. A. (2007). *Elementary and middle school mathematics: Teaching developmentally.* New York: Pearson Education.

---

[1]This was an earlier version of the case found in Smith, M. S., Silver, E. A., Stein, M. K., Henningsen, M. A., Boston, M., & Hughes, E. K. (2005).

Marjorie Henningsen lives in Beirut, Lebanon where she teaches mathematics education and directs the Science and Mathematics Education Center at the American University of Beirut. Dr. Henningsen completed her B.A. in Mathematics and Psychology at Benedictine College in Atchison, KS, followed by her M.Ed. and Ed.D. in Curriculum and Instruction (Mathematics) from the University of Pittsburgh. She has spent the past 18 years doing classroom research and designing and conducting teacher professional development in the U.S. and across the Middle East. In Fall 2007 she opened Wellspring Learning Community, a new international K-12 school in Beirut, Lebanon.

Steele, M. D.
AMTE Monograph 4
Cases in Mathematics Teacher Education: Tools for Developing Knowledge Needed for Teaching
©2008, pp. 57-72

6

Building Bridges: Cases as Catalysts for the Integration of Mathematical and Pedagogical Knowledge[1]

Michael D. Steele
Michigan State University

Cases can be used in a variety of ways with practicing teachers and teacher candidates to explore classroom practice and build knowledge needed for teaching. The richness of a case that portrays an authentic episode of teaching can afford teachers the opportunity to examine issues of mathematics content, pedagogical practices that support student learning, and the interactions between the two. In this chapter, I discuss a way of using cases in conjunction with the solving of a mathematical task that allows teachers to build these different facets of mathematical knowledge for teaching and strengthen the connections between them. Using records from a content-focused methods course for practicing teachers and teacher candidates, I examine how solving and discussing a mathematical task, then reading and analyzing a case of classroom practice featuring the same task, influenced teacher learning. This combination of exploring a mathematical task with the analysis of a related case afforded teachers the opportunity to integrate their mathematical, pedagogical, and pedagogical content knowledge in interesting ways.

What is the mathematical knowledge needed for teaching, and how might preservice and practicing teachers develop it? This question is at the heart of the work of mathematics teacher educators. As a field, mathematics education has few agreed-upon answers to this question; however, there are several characteristics of the knowledge base that are taken as shared. The knowledge base for teaching mathematics is complex and nuanced, encompassing mathematical knowledge, pedagogical knowledge, and more specialized knowledge for teaching that resides at the intersection of the two (e.g., Ball, Bass, & Hill, 2004; Sherin, 2002; Shulman, 1986). To teach mathematics effectively, teachers need to possess deep and well-connected understandings in each of these domains (Hill, Rowan, & Ball, 2005). Developing and connecting this knowledge can be challenging, in part because the structures for reasoning in the domains of mathematics, pedagogy, and specialized pedagogical content knowledge are different (Steele, 2005).

The complex nature of the knowledge base for teaching mathematics positions cases as promising learning tools for teacher candidates and practicing teachers. Cases that portray authentic episodes of teaching are rich representations of the complexity of instructional practice. Cases also provide insights into the links between pedagogical decision-making and student learning, particularly cases that include a teacher's thought processes through first-person narrative or pre- and post-lesson interviews. Cases offer practicing teachers and teacher candidates the opportunity to explore mathematical knowledge needed for teaching as they examine issues of mathematics content through the mathematical task featured in the case and the understandings that students portrayed in the case do (or do not) construct. Cases also highlight the pedagogical decisions made in teaching the lesson and often some or all of the reasoning behind these decisions. These decisions may not be purely pedagogical in nature, as they serve to advance the mathematics of the classroom. Thus, cases provide opportunities to examine and connect issues of mathematics content, pedagogical practices, and pedagogical content knowledge.

Content-focused Methods Courses: A Rich Venue for Case Analysis

Several researchers (e.g., Shulman, 1986, 1987; Smith, 2001; Sykes & Bird, 1992) have called for teacher educators to integrate cases into teacher preparation and professional development programs. The ASTEROID (A Study in Teacher Education: Research on Instructional Design) Project sought to develop three content-focused mathematics methods courses for both teacher candidates and practicing teachers, each centered on one of three mathematical strands central to middle grades mathematics – proportional reasoning, algebra as the study of patterns and functions, and geometry and measurement. Each course was built around a set of four narrative cases (see Smith, Silver, & Stein, 2005); each case portrays events in an urban middle grades classroom as a teacher engages students in work on a cognitively demanding mathematical task (Stein, Smith, Henningsen, & Silver, 2000).

Each course was designed to engage a mixed group of practicing teachers and teacher candidates at the elementary and secondary levels in the exploration of mathematics content, the analysis of student-centered pedagogical practices, and reflection on their own teaching practice. The courses were taught at a large public university in the eastern United States during a 6-week summer session, twice a week for 3 hours each session. The design of the courses was grounded in the tradition of practice-based teacher education (e.g., Ball & Cohen, 1999; Smith, 2001), in which the authentic work of teaching is examined as a tool for learning and reflection.

To take advantage of the potential of a case to examine issues of mathematics content, pedagogical practices, and pedagogical content knowledge, those participating in a course were asked to explore and discuss the middle grades mathematics task featured in the case first. It was anticipated that through the deep analysis and discussion of the mathematical task, teachers would have new insight into the teaching and learning described in the case. Moreover, engaging in the mathematics afforded participants the opportunity to make connections between the mathematical and pedagogical issues portrayed in the case. Following analysis of the case, teachers connected their learning to practice through activities such as analyzing student work or planning a lesson with similar mathematical content.

This chapter investigates how engaging practicing teachers and teacher candidates in solving and discussing a mathematical task, then analyzing a narrative case of teaching, offers unique opportunities to develop mathematical knowledge for teaching. Specifically, I explore the ways in which solving and discussing an authentic middle-grades mathematical task affects the discussion and analysis of a narrative case featuring that same task. The case described is taken from the ASTEROID course focused on the mathematical content of geometry and measurement. (For additional information on the course, the cases, and what teachers learned from the course, see Smith, Silver, & Stein (2005) and Steele (2006).)

The Case of Keith Campbell

The Case of Keith Campbell (Smith, Silver, & Stein, 2005) describes an experienced teacher and his 7th grade class engaged in an extended investigation of surface area and volume of rectangular prisms. Mr. Campbell's teaching can be characterized as in transition from a directive style of teaching to a student-centered pedagogy in which inquiry, communication, and high-level thinking are valued. The case, written in the teacher's voice, provides contextual background about Mr. Campbell's school, class, and previous experiences. The case provides a summary of the previous class discussions and description of the previous night's assignment, which is to be the discussion topic for the day's class meeting. In the previous lesson, students had constructed rectangular prisms with volumes of 1, 2, 3, 4, and 6 cubic units using interlocking cubes ("moon gems" in the context of the task). Students recorded the dimensions and surface area (cost of the moon gem packaging) of the prisms in a table and were to extend the table for a volume of 8 gems and make observations about the table for homework (see Figure 1).

The lesson featured in the case begins with students sharing their responses for 8 gems, with students presenting their answers, drawing diagrams and constructing explanations, questioning one another, and being questioned by Mr. Campbell. The exploration was intended for students to develop a conceptual understanding of volume and surface area based on their intuitive notions about the quantities. Students debate a number of key mathematical issues, including which arrangements of dimensions count as unique, various methods for determining the surface area, and relationships between the factors of 8 and the dimensions of the rectangular prisms. Students then engage in the lesson's primary task: exploring possible package dimensions for 9, 10, 11, and 12 gems. The balance of the case shows Mr. Campbell monitoring students' explorations and facilitating a discussion of the results for 9, 10, 11, and 12 gems, and moving the class towards the development of the formula *length* × *width* × *height* = *volume* for rectangular prisms.

Homework: Using the data in the table we created in class, explore the various ways of packaging 8 gems and add this information to the chart. Make observations and conjectures about the data we have recorded so far.

Volume in cubic units (# of gems)	Dimensions			Surface area in square units (cost)
	Front Edge	Side Edge	Height	
1	1	1	1	6
2	2	1	1	10
3	3	1	1	14
4	2	2	1	16
	4	1	1	18
6	3	2	1	22
	6	1	1	26

Figure 1. The Moon Gems task in *The Case of Keith Campbell*

Data Analysis

The Case of Keith Campbell was the third of four narrative cases discussed in the content-focused methods course related to geometry and measurement and served as a starting point for discussions of the relationships between dimension, surface area, and volume. Videotapes of course meetings, written artifacts, and assignments relevant to *The Case of Keith Campbell* were examined to determine the mathematical ideas raised by practicing teachers and teacher candidates in the public discussion. Discussions and assignments related to the analysis of the case were then examined for instances in which mathematical ideas raised during work on the task were also discussed in the case analysis. All teacher names in the excerpts that follow are pseudonyms.

Results

Teachers began work on *The Case of Keith Campbell* with the exploration of the *Arranging Cubes* task (Figure 2), a mathematical task similar to the primary task from Keith Campbell's lesson as described in the case. Work on the *Arranging Cubes* task and the analysis of the case took place across Classes 7 and 8 of the 12-session geometry and measurement course. Teachers worked in small groups to solve the task during Class 7, followed by a brief discussion of the solutions at the end of class. Teachers were asked to read *The Case of Keith Campbell* between Classes 7 and 8. Class 8 continued the discussion of the math task and the "Consider" questions (see Figure 2), followed by discussion of the narrative case during which teachers identified the mathematics ideas students learned and the moves Mr. Campbell made that supported or inhibited those ideas. Table 1 shows the sequence of events related to the case and the durations of the whole-class discussions related to each activity.

<u>Solve.</u>
Find all of the ways that following fixed numbers of cubes can be arranged into rectangular prisms: 8, 9, 10, 11, and 12. For each fixed number of cubes, sketch the rectangular prisms you create, and record their dimensions, volume, and surface area. You may want to organize your information into a table.
<u>Consider</u>

1. For each fixed number of cubes, how do you know that you have found all the rectangular prisms that can be constructed?
2. Explain why the formulas $SA = 2lw + 2lh + 2wh$ and $V = l \times w \times h$ can be used to determine the surface area and volume (respectively) of a rectangular prism.
3. For each of the fixed number of cubes, compare the prism with the greatest surface area to the one with the least surface area. Make observations about the characteristics of these prisms that appear to affect their surface area. Would the observations you made continue to be true for any set of rectangular prisms that share a constant volume?

Figure 2. The *Arranging Cubes* task
Adapted from *Middle Grades Mathematics Project: The Mouse and the Elephant: Measuring Growth Activity 4 and 5* by Janet Schroyer and William Fitzgerald, © 1986 by Pearson Education, Inc., or its affiliate(s). Used by permission.

Table 1. Class Discussions Related to *The Case of Keith Campbell*

Class 7	Class 8
Discussion of solutions to the *Arranging Cubes* task (10 min.)	Continued discussion of solutions to the *Arranging Cubes* task (20 min.)
	Discussion of the "Consider" questions from the *Arranging Cubes* task (15 min.)
	Discussion and analysis of *The Case of Keith Campbell* (45 min.)

Through the examination of the videotaped discussions around the mathematical task (Classes 7 and 8) and the analysis of the case (Class 8), three ideas were identified that were evident in both the discussion of the task and the analysis of the case: developing meaning for surface area through tasks and tools; creating meaning for formulas; and aligning tasks and goals. The sections that follow describe these three ideas; for each, excerpts from the discussion of the mathematical task and excerpts from the discussion of the case are provided.

Developing Meaning for Surface Area through Tasks and Tools

The task. The *Arranging Cubes* task can be classified as a mathematical task of high cognitive demand (Stein, Grover, & Henningsen, 1996). Specifically, the task as written does not provide an explicit or implicit solution path, and affords the use of non-algorithmic thinking and multiple representations. In launching the task in the geometry and measurement course, the instructor provided a variety of tools for the participants to use in making sense of the task: interlocking centimeter cubes, grid paper, and straight edges. During the sharing of solutions, the instructor solicited multiple methods for finding surface area, and pressed teachers to connect their numerical calculations to meaning through the use of diagrams and manipulatives. Excerpt 1 illustrates the ways in which solutions were shared for the surface area of 8-cube prisms.

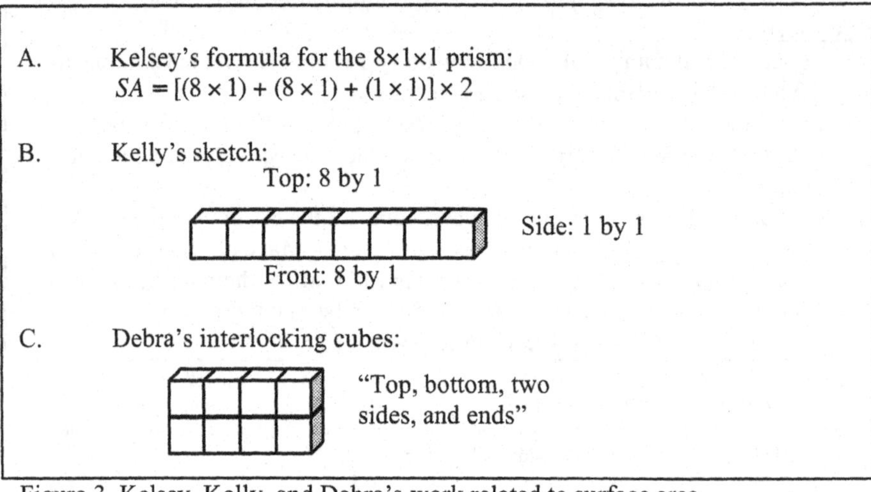

Figure 3. Kelsey, Kelly, and Debra's work related to surface area

Instructor: So for 8 cubes, for the 8 by 1 by 1 prism, would someone like to give us the volume and surface area for that? Kelsey, go ahead.

Kelsey: The volume is 8, and the surface area is 34.

Instructor: Can you tell me how you came up with 34?

Kelsey: I tend to just find the area — I wasn't even visualizing at that point. I guess I just knew that the 8 had to go with the 1, the 8 had to go with the other 1, and the 1 had to go with the 1, so that's 8, 8 and 1 which is 17. [See Figure 3a]

Instructor: So the 8 had to go with the 1—

Kelsey: Right. There had to be an 8 by 1 surface, there had to be another 8 by 1 surface, there had to be a 1 by 1 surface. And then I just doubled it, to take care of the other surfaces…

Instructor: So Kelsey approached it, 8 times 1, [pause] 8 times the other 1, and the 1 times 1, and then doubled it, and why did you

	double it?
Kelsey:	Because I knew that there were 6 surfaces and I knew, obviously, the surfaces were congruent.
Instructor:	Okay. Did anyone approach the surface area of this one a different way? Kelly?
Kelly:	I didn't look at the numbers as much as I looked at the sketch… And so I looked at each sketch so I knew each (set of) two dimensions. And then I did what Kelsey did, and I doubled it. I multiplied the 2 dimensions to get the area of that face, and then I doubled.
Instructor:	So in this case our sketch looks something like this. [See Figure 3b] So when you're saying the two dimensions you got, Kelly…
Kelly:	So if I looked at the top of that prism, that was an 8 by 1 surface.
Instructor:	Okay, so this is 8, and this is 1 there [points to Fig. 3b].
Kelly:	So I knew that area was 8, and then I doubled it. And then I did that with the side [points to diagram], which was 1 by 1, and then I did it with the front face, which is 8 by 1.
Instructor:	Okay. Any other ways of going about it? Debra?
Debra:	I looked at the top like she said and saw that it was 8 by 1 so I knew that the area was 8, and then I just found that all the way around 4 times. So 32, and then the 2 ends.
Instructor:	So 8 here, right? [points to front face] And then 1, 2, and then 3 and 4 that we can't see. [points to four 8-by-1 faces] Did that work for all of your prisms?
Debra:	No.
Instructor:	Give me one that it didn't work for.
Debra:	For 8 cubes… 2 by 4.
Instructor:	4 by 2 by 1? And what's the volume of that one?
Debra:	8.
Instructor:	And the surface area?
Debra:	28.
Instructor:	So how did you figure out the 28?
Debra:	I had to do it like, Kelly did. So like the top and the bottom and the two sides, and the ends. [Debra holds up her cube prism – see Fig. 3c]
Instructor:	[sketches the prism in Fig. 3c] So you looked at the top and bottom, and the two sides, and the ends. So can you say what those were based on my sketch?
Debra:	2 by 4
Instructor:	2 and a 4, ok.
Debra:	And then the top is 4 (by 1)…
Instructor:	So, so Debra went about it, 4 times 1, 2 times 4, 1 times 2, and then all that times 2, I think? [Debra nods yes] Similar to what Kelsey was talking about and what Kelly was

talking about based on the diagram.

Excerpt 1, Class 8

In Excerpt 1, three different teachers use words, numbers, cube models, and diagrams to explain the methods they used to find the surface area of the rectangular prisms. Teachers were able to approach this task using methods with which they were comfortable: Kelsey relied on the numbers, Kelly created and used diagrams, and Debra used the cube models. Later in the discussion, the instructor asked teachers how the cubes helped them in their work on the task.

Instructor: How did having the cubes in front of you help you, in terms of answering the questions that I had asked you to answer?

Bridget: It helped me make sure that I was making each prism I could with the cubes. Because if I went on the drawing I probably would have left some of them out without even realizing it, but since I had the cubes to lay out, I could see each way to make the prisms.

Instructor: So it helped you make the combinations, and kind of organize how you made them…

Ed: It helped to see that when you had a shape that looked like it was different, you could actually pick it up and physically twist it, and it showed you that it was the same dimensions. Like a $4 \times 2 \times 1$ is the same as a $1 \times 4 \times 2$. It was just easier to see it that way.

Instructor: So it helped with that issue of duplication and making that argument – in fact, I think when we made that argument someone held one up and rotated it. Debbie?

Debbie: It was easier to find the volume, because you knew you had 8 cubes, and you used those 8 cubes to make the rectangular prism. So you didn't actually have to calculate anything, you knew you had 8—

Nick: I didn't use the cubes. I just started by sketching everything.

Instructor: [Nick] just went right to drawing. A lot of people went right to the cubes. But you didn't have to use the cubes, to be successful. So Nick took another entry point into the task that didn't involve the cubes, and was just as successful.

Excerpt 2, Class 8

In this conversation, teachers not only explain how the cubes helped them approach the task, but also suggest that the cubes were not the sole point of entry for making sense of surface area. This notion of multiple entry points was also salient during the discussion of the case.

The case. In discussing *The Case of Keith Campbell*, teachers were asked to identify mathematical ideas that Mr. Campbell's students were learning, and then to identify pedagogical moves that either supported or inhibited his students' learning of those mathematical ideas. One of the first mathematical ideas identified was *surface area*, a concept that was central to the *Arranging Cubes* task. In Excerpt 3, Bridget raises the issue of affording entry to the task as critical in supporting students' learning about surface area.

Instructor: Other support or inhibit factors for surface area? Bridget?

Bridget: I think it was [paragraph] 25, he was allowing the students to explore in whatever way they were comfortable with, he didn't force them to use the manipulatives, they could sketch if they wanted or some of them went just right to the table. And I thought that was a support factor, allowing the kids to look at solving a problem (their own way).

Instructor: So what Bridget was saying is this idea that he let students enter the task at different points, helped students go where they were comfortable. And certainly this idea of multiple entry points is something that we've discussed in the contexts of other tasks, and I think came up when we talked about how the cubes helped us make sense of the task I gave you.

Excerpt 3, Class 8

Bridget identifies the open-ended launch of the task and the use of manipulatives as supporting the learning of Mr. Campbell's students. This statement mirrors an idea that the class had previously cited as contributing to their *own* mathematical understanding during their work on the mathematical task. The instructor points out the link between the pedagogical move that Bridget identified in the case and the move that the practicing teachers and teacher candidates in the course had identified as supporting their own learning. This suggests that teachers' work on the mathematical task may have helped them to notice this pedagogical move in their analysis of the case.

Creating Meaning for Formulas

The task. During the sharing of solutions for the *Arranging Cubes* task before reading the case, Noelle shared a method for finding surface area that involved finding the perimeter of the prism, multiplying by the height, and adding in the area of the two bases. The instructor pressed Noelle to connect her method to the structure of the prism in order to help others understand the method and to connect the method to the meaning of surface area. Excerpt 4 shows part of the 5-minute exchange.

Instructor: So Debra went about it, 4 times 1, 2 times 4, 1 times 2, and then all that times 2, I think? Similar to what Kelsey was talking about and what Kelly was talking about based on the diagram. Now Noelle, explain what you did.

Noelle: I did what she did, but I factored out, the 2…She [Kelly] found the area of each side. For the 1 times 2, 1 times 2, 4 times 2 two times. So I just pulled the 2 out.

Instructor:	So tell me the mathematical expression that you were-
Noelle:	$2 \times (1 + 1 + 4 + 4)$
Instructor:	Alright. So where-
Noelle:	That's not the whole surface area.
Instructor:	Okay, so what is it.
Noelle:	It's the area around. It's the sides – it's not the tops.
Instructor:	Come show me so I've got it correct.
Noelle:	[Points to four lateral sides – see Fig. 4a]
Instructor:	So when we think about…
Kelly:	It's like she unfolded the sides, and found the area of one large rectangle.
Instructor:	So if we unfold this, [see Fig. 4b] this is the front – let me put a box around this so we know it's the front. So there's the front, there are the two sides kind of like wings, let's put the back out here… so, now. Noelle, now that I've drawn this, can you tell me where the two 1s and two 4s came from?
Noelle:	The 2 is the height of each face, and the 1 is the length of the two, sides.
Instructor:	And the 4s?
Noelle:	The 4s go to the front.

Excerpt 4, Class 8

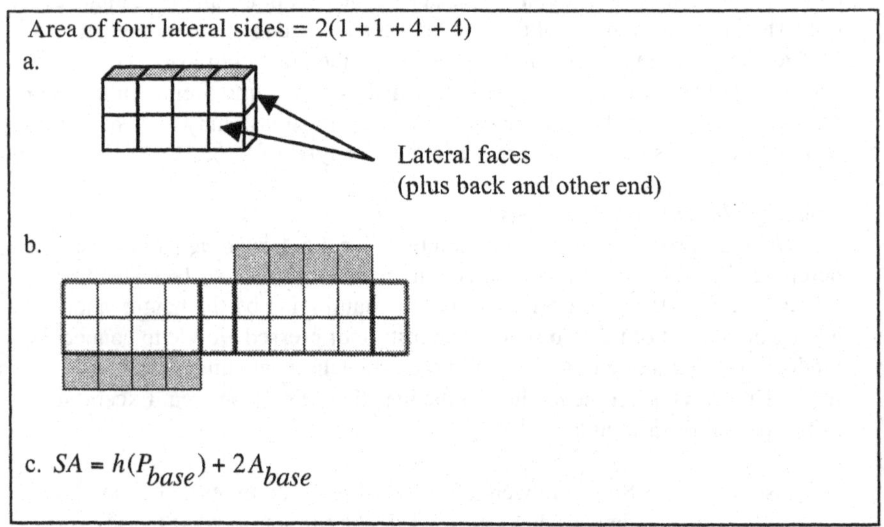

Figure 4. Noelle's method for finding surface area

Noelle's talk and gestures show her connecting her method of finding the surface area (using the perimeter of the base) with the diagram, and relating her method to Kelly's method. This might suggest that she has a strong understanding of how her calculation relates to the visual representation of the prism. However, a comment by Noelle later in the class casts this contribution in a new light:

Noelle:	This is the opposite of it, I have to use the formula. I have a really hard time doing that [visualizing] I can't see a three dimensions [*sic*] figure, cut it, and open it up. So if I know the formula, for every single prism, I know I'm going to get it right. I couldn't see that it made a rectangle. This is easier but if I couldn't see it, I might mess it up. It's kind of opposite of what you want to hear [laugh]. But memorizing a formula kind of helps me.
[instructor asks Noelle to explain her formula again]	
Noelle:	So you're finding… the area of each side and adding it together. So when you find the area of each face-
Instructor:	So why does that work for any prism?
Noelle:	Because any prism has… I don't know if I can explain it….
Instructor:	So say again the formula?
Noelle:	Height, times the perimeter of the base, plus twice the area of the base. [see Fig. 4c]

Excerpt 5, Class 8

Noelle's admission that she tends to rely on the formula and has difficulty visualizing suggests that without the press of the instructor and Kelly's support, Noelle may not have connected her surface area calculation to the attributes of the rectangular prism. Rather, she might have just identified the necessary numbers from the prism to substitute into her memorized formula and used that method to calculate the surface area.

The case. In *The Case of Keith Campbell,* Mr. Campbell holds as a goal for his students to develop the formula for the volume of a rectangular prism, *Volume = length × width × height.* Mr. Campbell's abrupt introduction of the formula at the close of the case was a significant point of discussion for the teacher candidates and practicing teachers in the course. In this short excerpt, teachers in the course are discussing whether Mr. Campbell jumped to the formula too quickly to create meaning for students, and whether or not the work done by the class constituted a proof that the formula was valid. Florence makes a comparison between the work of Mr. Campbell's class and Noelle's work on the task.

Florence:	There was a reason Noelle chose the numbers she chose, and they [Mr. Campbell's class] don't have reasons for these numbers at this point. [To be a proof,] you have to have reasons and um, justifications, they just have examples that work, with no justifications as to WHY those numbers work.

Excerpt 6, Class 8

Florence's comment suggests that the class' experience solving the task, and specifically Noelle's work in explaining her formula for surface area, influenced her analysis of the case. Up to this point in the case discussion, a number of teachers in the course had suggested that simply learning to apply a formula did not guarantee that students were grappling with the key understandings related to surface area. Florence here suggests that Noelle's use of the formula showed evidence of mathematical reasoning, while the use of the formula in the case did not.

Uma, a practicing teacher in the course, connects both the solution of the task and the case analysis experience to her own teaching in a short writing assignment in which course teachers were asked to identify the experiences students should have with respect to surface area and volume in the middle grades.

> *The Case of Keith Campbell* also relates the idea of students discovering their own connections to the concept of surface area. The students in his class are working to discover how dimensions of edge length relate to finding surface area. The more work we do in class and the more cases we look at, lead me to believe that students get more from an activity than they would from a lecture. Students working through their own reasoning and discovery seem to make more and longer lasting connections to the mathematical concepts they are learning.
>
> Excerpt 7, Learning Log 4, Uma

In sum, these four excerpts suggest that the work on the *Arranging Cubes* task had an influence on the ways in which teachers in the course viewed the teaching and learning portrayed in *The Case of Keith Campbell*, in particular with respect to students making meaning for surface area. Moreover, Excerpt 7 shows a teacher connecting experiences with both the mathematical task and the case analysis to her own teaching practice.

Aligning Tasks and Goals

The task. The *Arranging Cubes* task ostensibly allows for the exploration of the concepts of surface area and volume. The task requests the surface area and volume for each prism created, along with the dimensions, and in the "Consider" questions the formula for volume must be explained. As demonstrated by Excerpts 1 and 4, the task affords a wide variety of solution methods with respect to surface area, and has the capacity to illuminate the benefits and limitations of different ways of finding the surface area for different prism configurations. Although the task does involve volume, the range of understandings the task affords with respect to volume is limited. During the work on the task, teachers in the course stopped to consider volume only briefly, in the exchange documented in Excerpt 8.

Instructor: So how about the $2 \times 2 \times 2$, for 8 cubes? What's our volume, what's our surface area?

(inaudible teacher response)

Instructor:	8 and… 24. So why are their [volumes] all the same? Florence?
Florence:	Any way you arrange them, 8 cubes makes up the figure. So it doesn't matter how you arrange them, there are 8 cubic units that are making up that figure any way you put them.
Instructor:	So how do the cubes relate to volume?
Florence:	Volume is how much is inside something, how much fills it, so if we had an irregular shape it would still take those 8 cubic units to fill it.
Instructor:	So even if I put something together that looked like, let's say, [draws 8 cubes in an arrangement that wasn't a rectangular prism] that, it's still 8 cubic units.

Excerpt 8, Class 8

The case. In *The Case of Keith Campbell*, Mr. Campbell cites the development of a formula for volume of a rectangular prism as one of his two primary goals for the lesson. But in analyzing the case, the practicing teachers and teacher candidates noted that the task may not have afforded students rich opportunities to consider the *nature* of volume. In discussing the pedagogical moves that supported or inhibited student learning of volume, teachers in the course stated that they were not clear about what students understood about volume in the case, and wondered whether the development of the formula in the case was helpful or harmful. Uri, a preservice teacher, summed up these concerns during a discussion of what could be taken from the case.

Instructor:	So, what lessons about teaching in general can we learn from this case? What can we take away from this? Uri?
Uri:	I guess kind of what Florence and Noelle were saying about, how this task didn't really get at the idea of volume that well… [inaudible speech]
Instructor:	So it goes back to considering what your task should be to accomplish the goal you're setting out to accomplish.

Excerpt 9, Class 8

In solving and exploring the mathematical task, the participants in the course had the opportunity to experience first-hand how the task did or did not afford insight into surface area and volume. The notion that the task may not have been a rich opportunity to develop a conceptual understanding of volume was evident in the discussion and analysis of the case. If teachers had simply read over the task prior to engaging with the case, this aspect of the task may not have been as salient.

Discussion

Cases that are authentic portraits of classroom practice and teacher decision-making are well positioned as vehicles for teachers to develop forms of pedagogical content knowledge that are deep, nuanced, and well connected. By engaging teachers and teacher candidates in activities that feature thoughtful work on a mathematical task followed by the analysis of a case featuring that task, teachers appear more likely to integrate the mathematical and pedagogical issues at play in the case. The excerpts presented in the previous sections show teachers bringing the mathematical understandings discussed during solving of the task to bear on their analysis of the narrative case. In analyzing the case, teachers did not simply identify pedagogical issues; they noted moves made by Mr. Campbell that were tied to students' developing understandings of the mathematics. This sequence of solving a task and analyzing a case thus affords teachers opportunities to develop important mathematical knowledge for teaching – in particular, the pedagogical content knowledge that links teachers' mathematical knowledge to their pedagogical knowledge.

So what might practicing teachers and teacher candidates have taken away from their experiences with the *Arranging Cubes* task and *The Case of Keith Campbell*? The excerpts from the task discussions suggest that rich, detailed explanations for finding the surface area of rectangular prisms were available to teachers. The excerpts from the case discussions show teachers using these mathematical understandings as a means to contextualize the teaching decisions Mr. Campbell makes and the learning of his students. In this way, teachers had the opportunity to integrate their knowledge of mathematics with their knowledge of pedagogy and create a more powerful learning experience than either activity might have afforded individually.

Ultimately, it is hoped that this knowledge has an impact on teachers' classroom practice. In reflecting on the Keith Campbell activities, some teachers in the course made statements that suggest a changing view on their own practice.

> Also, I remember that Florence kept pointing out that Campbell never discusses the cube as a unit that tells how much it would take to "fill the box" and can tell how much it takes to fill other shapes. Her point got me to thinking about all those lessons I have shoved in my filing cabinets that use non-standard units such as rice grains or beans to help kids explore volume. I have always skipped those lessons because I felt that Campbell's lesson addressed all the same learning objectives but goes one step farther by arriving at the $V = lwh$ formula… But now I'm thinking that the rice grains lesson does a lot more for kids than I thought it did before.
>
> Excerpt 10, Learning Log 4, Kelsey

> I have to admit that this year I taught two Geometry classes in which I had discussed various 3-dimensional shapes with my students and went over formulas, and that was the extent to surface area and volume. Reason being

is that this unit fell at the end of the school year and we had to touch on both measurements. In agreement with Campbell... I truly have the same opinion that my students have little true understanding of what surface area and volume entail.

Excerpt 11, Learning Log 3, Chuck

If one were to design a teacher education experience aimed at enhancing mathematical knowledge for teaching, how might cases be used effectively? The results discussed here suggest that work on a mathematical task and the analysis of a related case enhances the learning of mathematical, pedagogical, and pedagogical content knowledge. A course experience built around a particular mathematical focus (such as geometry and measurement), with tasks and cases that highlight different aspects of the mathematical content, potentially positions teachers on a learning trajectory that builds mathematical and pedagogical knowledge in ways that are integrated and useful in the work of teaching.

References

Ball, D. L., Bass, H., & Hill, H. C. (2004, January). Knowing and using mathematical knowledge in teaching: Learning what matters. Paper presented at the meeting of the South African Association of Mathematics, Science, and Technology Education, Cape Town, South Africa.

Ball, D. L., & Cohen, D. (1999). Developing practice, developing practitioners: Toward a practice based theory of professional education. In G. Sykes & L. Darling-Hammond (Eds.), *Teaching as the learning profession: Handbook of policy and practice* (pp. 3-32). San Francisco: Jossey-Bass.

Hill, H., Rowan, B., & Ball, D. L. (2005). Effects of teachers' mathematical knowledge for teaching on student achievement. *American Educational Research Journal, 42*(2), 371-406.

Sherin, M. G. (2002). When teaching becomes learning. *Cognition and Instruction, 20*(2), 119-150.

Shulman, L. S. (1986). Those who understand: knowledge growth in teaching. *Educational Researcher, 15*(2), 4-14.

Shulman, L. S. (1987). Knowledge and teaching: Foundations of the new reform. *Harvard Educational Review, 57*(1), 1-22.

Smith, M. S. (2001). *Practice-based professional development for teachers of mathematics*. Reston, VA: National Council of Teachers of Mathematics.

Smith, M. S., Silver, E. A., & Stein, M. K. (2005). *Improving instruction in geometry and measurement: Using cases to transform mathematics teaching and learning* (Volume 3). New York: Teachers College Press.

Steele, M. D. (2005). Comparing knowledge bases and reasoning structures in discussions of mathematics and pedagogy. *Journal of Mathematics Teacher Education, 8*(4), 291-328.

Steele, M. D. (2006). *Middle grades geometry and measurement: Examining change in knowledge needed for teaching through a practice-based teacher education experience.* Unpublished doctoral dissertation, University of Pittsburgh.

Stein, M. K., Grover, B. W., & Henningsen, M. A. (1996). Building student capacity for mathematical thinking and reasoning: An analysis of mathematical tasks used in reform classrooms. *American Educational Research Journal, 33*(2), 455-488.

Stein, M. K., Smith, M. S., Henningsen, M. A., & Silver, E. A. (2000). *Implementing standards-based mathematics instruction: A casebook for professional development.* New York: Teachers College Press.

Sykes, G., & Bird, T. (1992). Teacher education and the case idea. *Review of Research in Education, 18*, 457-521.

[1]This work was supported in part by a grant from the National Science Foundation (0101799) for the ASTEROID Project. Any opinions expressed herein are those of the author and do not necessarily represent the views of the Foundation.

Michael D. Steele is an Assistant Professor of Teacher Education at Michigan State University. His research interests include investigating mathematical knowledge for teaching, and specifically how practicing teachers and teacher candidates acquire this knowledge and make use of it in classroom practice. A former middle and high school math and science teacher, he currently teaches in the elementary and secondary teacher preparation programs at MSU. His work as a doctoral student at the University of Pittsburgh included the design, enactment, and study of content-focused methods courses for the NSF-funded ASTEROID Project, directed by Margaret S. Smith.

Hillen, A. F. & Hughes, E. K.
AMTE Monograph 4
Cases in Mathematics Teacher Education: Tools for Developing Knowledge Needed for Teaching
©2008, pp. 73-88

7

Developing Teachers' Abilities to Facilitate Meaningful Classroom Discourse Through Cases: The Case of Accountable Talk[1]

Amy F. Hillen
Robert Morris University

Elizabeth K. Hughes
Education Consultant

Facilitating mathematical discussions in which students engage in high-level thinking and reasoning about mathematics is a critical, yet daunting task for teachers. Using accountable talk is one way in which teachers can support students' learning of mathematics in discourse-oriented classrooms. In this chapter, we describe the use of The Case of Edith Hart in a teacher education course, and consider how teachers' participation in a set of activities centered on the case – specifically, a mathematical discussion in which teachers were held accountable, and reading and discussing a case in which the students were held accountable – provided opportunities to identify aspects of accountable talk. In particular, we draw upon evidence from the case discussion to illustrate the ways in which teachers identified accountable talk moves made by the case teacher, linked these moves to students' learning, and moved from the particulars of the case to developing generalizations about facilitating effective classroom discussions. We conclude with an analysis of the key role the case facilitator played in supporting teachers' learning and offer some suggestions for teacher educators who are facilitating teachers' work with cases.

Communication is a key component of classrooms in which students engage in high-level thinking and reasoning about mathematics (Carpenter & Lehrer, 1999; Hiebert et al., 1997; NCTM, 2000). In particular, Carpenter and Lehrer (1999) argue that "classrooms are discourse communities in which all students discuss alternative strategies or different ways of viewing important mathematical ideas" (p. 26). In such classrooms, teachers have a daunting task – managing the ideas of individual students and ensuring that the important mathematical ideas of the lesson are being made public, all the while helping students learn how to participate in a discourse-oriented classroom (Ball, 2001; Hiebert et al., 1997; Lampert, 2001; Leinhardt & Steele, 2005). Meeting these goals requires that teachers take an active role in facilitating mathematical discussions (Chazan & Ball, 1999). Because this role may be new to teachers, they need opportunities to develop a repertoire of ways to orchestrate productive mathematical discussions (Chazan & Ball, 1999).

One way in which teachers can support students' learning of mathematics in discourse-oriented classrooms is to establish a classroom norm called *accountable talk* – that is, talk that is accountable to the classroom community and to the discipline of mathematics, and seriously responds to and further develops what others in the group have said (Michaels et al., 2002; Resnick, 1999). Teachers hold students accountable by modeling appropriate forms of discussion, by questioning, and by facilitating discussions (Michaels et al., 2002). For example, teachers might press for clarification and explanation by asking students questions such as "What do you mean by…?" or "Can you say more about…?". Teachers might also ask questions that press students to justify their thinking or provide evidence for their claims, such as "Why does…?" or "How do you know…?". In addition, teachers might hold students accountable by restating or revoicing students' contributions (Michaels et al., 2002; O'Connor & Michaels, 1993, 1996). By engaging in mathematical discussions in which they are held accountable, students also learn how to participate in mathematical discourse. In addition, students come to "expect that the teacher and their peers will want explanations as to why their conjectures and conclusions make sense and why a procedure they have used is valid for the given problem" (Carpenter & Lehrer, 1999, p. 26), and carry out accountable talk moves themselves (Michaels et al., 2002).

One promising approach for developing teachers' understandings about key pedagogical ideas, such as facilitating meaningful classroom discourse through accountable talk, is through the study of cases (Barnett, 1991, 1998; Merseth & Lacey, 1993; Shulman, 1996). In particular, some cases present detailed accounts of classroom discourse, and as such, provide opportunities for teachers to identify ways in which the teacher featured in the case supports students' learning. As teachers consider how the moves made by the case teacher appear to impact students' learning, they can also begin to develop generalizations about facilitating effective classroom discussions that are applicable beyond a particular case.

In this chapter, we describe the use of *The Case of Edith Hart* (Smith, Silver, & Stein, 2005) with preservice and practicing teachers enrolled in a mathematics methods course and consider how participating in a set of activities centered on the case – specifically, a mathematical discussion in which teachers were held accountable, and reading and discussing a case in which the students were held accountable – helped teachers identify aspects of accountable talk. We begin with a description of the context in which the case was used.

Setting the Context

The episode we describe in this chapter took place in a masters-level methods course that focused on algebra in the middle grades[2]. The course was held in a six-week term during the summer, and met twice a week for approximately 3 hours each session. The 21 teachers enrolled in the course varied with respect to teaching experience (including both preservice and practicing teachers) and certification (including elementary, secondary, and special education teachers).

The goal of the course was to help teachers construct (or reconstruct) their own understandings about algebra and functional relationships and to develop their capacity for providing meaningful learning opportunities for the students with whom they work. Towards that end, the course was *practice-based* (Ball & Cohen, 1999; Smith, 2001) in the sense that activities were grounded in authentic records of practice, such as student work or episodes of teaching depicted in written or video form.

A written, narrative case, *The Case of Edith Hart*, was selected as a part of the course for several reasons: 1) the task featured in the case, Cal's Dinner Card Deals (shown in Figure 1), is presented as a graph (rather than as equations) – a unique starting point for most teachers, as well as for the students in the case; 2) students in the case engaged in high-level thinking while laying the groundwork for making sense of linear functions; and 3) students' learning could be linked to Ms. Hart's practices and moves – many of which could be classified as accountable talk. Thus, the case had the potential to help teachers in the course consider particular practices and moves that support students' learning.

The Case of Edith Hart

The lesson depicted in *The Case of Edith Hart* features eighth-grade students exploring Cal's Dinner Card Deals (shown in Figure 1), a task that presents data about three different dinner plans in graphical form, and asks students to determine which plan is best by making sense of particular features of the graph (e.g., slope, y-intercept, points of intersection) in the dinner plan context. Ms. Hart's goal for the lesson is for students to reason about real-world contexts as they write formulas that will describe the linear relationship between the number of dinners purchased and the total cost. Although she does not plan to discuss the idea of slope formally, Ms. Hart hopes that students will notice the relationship between the steepness in the graph and the formulas of the dinner plans.

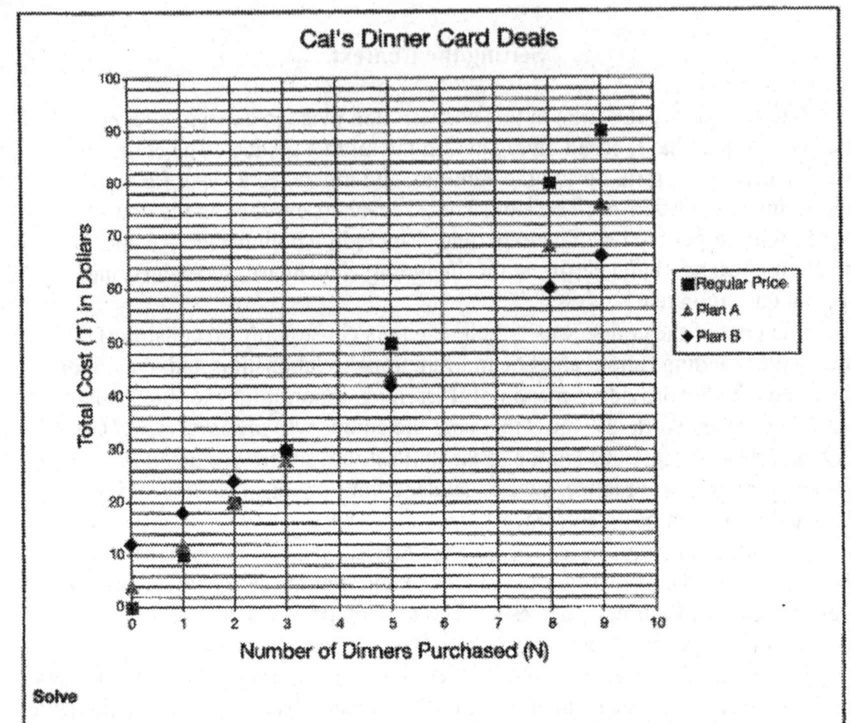

Solve

The graphs show data for three dinner plans. Make observations about each of the graphs. Find a way to determine the cost for any number of meals purchased on each dinner plan and explain your reasoning. Which dinner plan is best?

Consider

1. If you only had $5 to spend on food for the week, which dinner plan should you buy? How did you decide?
2. Does it make sense to connect the points in the dinner plan graphs? Why or why not?
3. Describe another situation besides dinner card plans that would generate a formula and a graph similar to the graph of Plan A or Plan B. Describe another situation that would give you a formula and a graph similar to the Regular Price Plan.

Figure 1. Cal's Dinner Card Deals (the task featured in *The Case of Edith Hart*)

Ms. Hart begins the lesson by asking students to work individually and make observations about the graphs. Students are then asked to share their observations with their small groups and to make some new observations. After about 15 minutes, students share their observations publicly in a whole group discussion. As students share their ideas, Ms. Hart repeats their observations and explanations and reminds students to record them in their notebooks.

After students exhaust a public sharing of observations, they continue working in small groups and develop different ways of determining the formulas that would generate each of the graphs: $8N + 4 = T$ for Plan A, $6N + 12 = T$ for Plan B, and $10N = T$ for the Regular Price Plan. After these formulas have been recorded on the overhead, several students make observations that lay the groundwork for discussing slope more formally in subsequent lessons (e.g., "the bigger the number being multiplied by *N*, the steeper the graph" (Smith, Silver, & Stein, 2005, p. 58)). During this part of the lesson, Ms. Hart presses students for further clarification of their conjectures. In addition, she holds all students in the class accountable for seriously considering one student's conjecture by repeating the conjecture and telling students to record it in their notebooks and to "think about whether they agreed with this statement" (Smith, Silver, & Stein, 2005, p. 58).

Thus, during the lesson featured in the case, Ms. Hart holds students accountable for clearly explaining their thinking and for listening carefully to others so as to clarify their statements. In the next section, we consider how the teachers in the course were also held accountable as they extended the Cal's Dinner Card Deals task and discussed *The Case of Edith Hart.*

Teachers' Work on The Case of Edith Hart

The 21 teachers in the course engaged in a series of activities centered on *The Case of Edith Hart* during the fourth week of the course that entailed first solving Cal's Dinner Card Deals. For homework, teachers then read the case and identified specific mathematical ideas they thought the students in the case learned (or were in the process of learning).

During the following class, teachers engaged in an activity in which they extended the mathematical ideas in Cal's Dinner Card Deals. In particular, teachers were asked to consider the average cost per meal (i.e., the total cost divided by the number of meals) for each of the three dinner plans and graph the average cost per meal as a function of the number of meals for each plan (shown in Figure 2). As teachers discussed these relationships, one teacher noticed that while the relationship in the Regular Price Plan is constant, the average cost per meal decreases in Plans A and B – and that the average cost per meal decreases faster in Plan B. The instructor pressed teachers to justify *why* this was so, and teachers debated whether the activation fee (e.g., the 12 dollars in Plan B) or the cost per meal (e.g., the 6 dollars in Plan B) was causing the average cost per meal for Plan B to decrease faster than in Plan A. At one point during the discussion, teachers drew upon mathematical ideas such as tangent lines to argue that the activation fee is causing the faster decrease in Plan B, as shown in the following excerpt:

1 Melanie[3]: I guess I am going back to the question of why does Plan
2 B decrease faster? Isn't that because of the twelve?
3 Owen: Yeah, it's because of the twelve, because the twelve's
4 bigger. And if you looked at the tangent to those two
5 lines, the one on [Plan B] would be steeper because it is
6 decreasing at a larger increment each time because the
7 twelve- when you split the twelve, there's still a larger
8 amount than when you split the four.
9 Kelsey: You mean the tangent at any specific number of meals, is
10 what you're saying?
11 Owen: Yeah, if you take the tangent at any particular x value,
12 you'll have a steeper slope on the [Plan B] graph.
13 Bonnie: But that doesn't have anything to do with the twelve or
14 the four.
15 Owen: Yes it does. It has to do with the twelve.

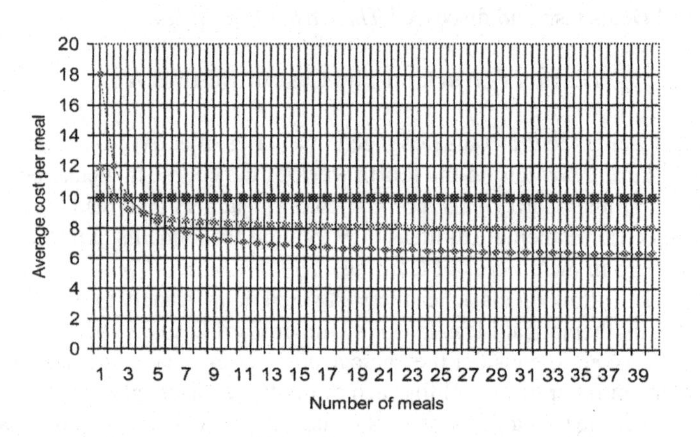

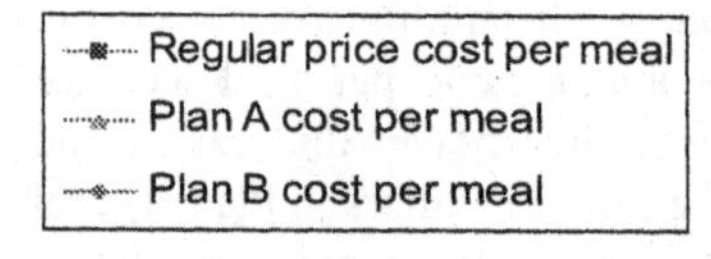

Figure 2. The average cost per meal graph that teachers created and discussed

The instructor then asked Owen to restate his point, and he noted that the equations for the average cost per meal were different from the original equations they had created during the prior class meeting. In order to clarify this point, the instructor asked the class to list the equations that correspond to the graphs shown in Figure 2 ($y = 8 + \frac{4}{x}$ for Plan A, $y = 6 + \frac{12}{x}$ for Plan B, and $y = 10$ for the Regular Price Plan). Kelsey then suggested alternative equations that correspond to the graphs shown in Figure 2. Throughout the discussion, the instructor used accountable talk moves to ensure that all teachers in the class were making sense of the ideas being discussed, as shown in the following excerpt:

Kelsey: Could I give an alternate equation that makes more sense to me? Cause I think it might make more sense to other people in the room. I just took the $8x + 4$ and then I just put that all over x, cause you're just taking the total price and dividing price by the number of meals, and that gives you the price per meal.

Instructor: And again if you divide both terms by x, you get six plus twelve over x or Owen's way, thinking about that this is the number of people who are splitting the activation fee. Okay. So, continue with what you were saying.

Kelsey: So, Owen was saying that, and that sort of clarifies the whole twelve versus six affecting the graph discussion, am I right?

Owen: Yeah.

Instructor: Can you say a little more about that?

Kelsey: Well, I think it's resolved, right? Everybody knows.

Instructor: I'm not convinced it's resolved. So I'd appreciate a little bit more language.

Kelsey: Well, the fact that it is the twelve that's being split up, not the six.

Instructor: Okay.

Kelsey: So, because you're splitting up a larger activation fee by a given number of meals, then that split-up amount is going to be larger per meal, or, larger per person. So then the decrease in the cost per meal would be a larger decrease because you're spreading out more—does that make sense?

Melanie: Yeah! She said the other ones shouldn't be decreasing faster because of the twelve, and that's when I got stuck. Then you guys went on for a few more minutes, and I was still thinking about what she said. I'm like, "Wait a minute. That doesn't sound right."

48	Instructor:	But what we've just demonstrated is how you can work
49		together as a group to try to make sense of something, and
50		when something comes out on the table that doesn't sit
51		right with you, then what you need to do is be diligent
52		about trying to make sense of it until it does, and in doing
53		that, it advances the entire group.

In this excerpt, the instructor pressed teachers for a rationale for why the average cost per meal decreases faster for Plan B than for Plan A. The instructor then supported teachers' work by revoicing (lines 22-25) and asking teachers to further clarify their thinking so that all teachers had an opportunity to make sense of the ideas being discussed (e.g., "I'm not convinced it's resolved. So I'd appreciate a little bit more language." [lines 32-33]). Following this discussion of the average cost per meal, teachers engaged in a discussion of *The Case of Edith Hart.*

The instructor began the case discussion by asking teachers to draw on their homework in which they had identified mathematical ideas they thought Ms. Hart's students learned or were in the process of learning. In particular, teachers worked in small groups to indicate how Ms. Hart facilitated or supported students' learning of each idea, and to cite evidence from the case to support their claims. After 20 minutes, the instructor called the groups together and asked the class to create a whole-class list. The whole-class list included both ideas related to mathematical content (e.g., slope as steepness of the line and rate of change) and mathematical processes (e.g., the ability to communicate ideas and evaluate themselves and each other). The instructor then selected four ideas for the class to discuss further: 1) points of intersection; 2) slope; 3) y-intercept; and 4) communication. In the remainder of the chapter, we examine teachers' discussion about the ways in which Ms. Hart supported students' learning of points of intersection and slope to provide evidence that teachers identified accountable talk moves made by Ms. Hart, discussed how these moves supported students' learning of mathematics, and moved beyond the particulars of the case.

The discussion began with teachers identifying moves made by Ms. Hart that supported her students' learning of points of intersection. For example, Bert observed that when a student in the case said he noticed that the graphs seem to cross at certain places, Ms. Hart "tried to put it into context, to see what it meant, and she repeated it." Bert noted, "repeating it was one way to enhance students' learning, they heard it again." Carl added to Bert's contribution by noting that Ms. Hart pressed for a generalization when she asked what it means in general when the graphs share a point, or intersect. Teachers continued to discuss how these moves served to support students' learning. Bruce then noticed a similarity between Ms. Hart's and the instructor's practice, as shown in the following excerpt:

Instructor: Bruce? You were going to add something?

Bruce: Well, I think it's similar to what you were doing earlier too, that you keep asking for clarification, until you get at what you're trying— it's almost like you're waiting for a specific sentence to be said, or different points of view to be brought to bear.

Instructor: Okay.

Bruce: When we were talking about the tangents [of Plan A and Plan B], you asked someone to repeat what he said, in your own words.

Instructor: Mhm.

Bruce: And so you kind of do that around the room. Edith Hart does that, a little bit, too, until she's satisfied that everyone has their own, maybe their own meaning, or they've said it in a way that she wants them to remember it maybe. Like what intercepts really are.

Instructor: You're identifying something that's really important, both in this lesson, and maybe bigger than that, pedagogically. So why did I do that? Why do I do that? Why does she do that, go around the room? Is it trying to get people to say something in a particular way or is there something else going on there? Lucy?

Lucy: She's asking a question, she needs to ask to pull mathematical ideas from the students, rather than providing them with the information she wants them to know.

Instructor: Okay.

Melanie: I was thinking that, maybe she does it because if a student can articulate an idea on their own, in a complete sentence, then you really are sure they really understand it, rather than the teacher articulating it first, and having them write it down, and memorize it or something. So if she asks them to say what they mean, and then someone else repeats it, then you really know they understand it.

Instructor: Well, that's consistent with the point that you [Lucy] were making. Ok, Bruce?

Bruce: I think, a lot of times, one student will say something, and it'll make sense to them, and maybe even the teacher knows that the student gets it. But in our example, we were talking about tangents of lines, and for some people, that might have been something that they learned years ago, or maybe they never even learned it. So it required someone else to explain what someone else had just said, and keep doing that, until everyone understands, and the way you do that is- I think you have to look out over your audience, and kind of see people's faces. And then,

100 someone says something about tangents, and lines, and
101 you said, "Well, what is a tangent?" So, you just have to
102 continue, to push, until you're satisfied as a teacher that
103 not just the person who raises their hand, but everyone
104 understands it to some degree.
105 Instructor: And there's another thing about what you're saying that's
106 really important, that is, knowing something about your
107 learners. Lisa, you had your hand up.
108 Lisa: Yeah, I was just going to say, also when you have
109 different students articulating the same concept, then,
110 other students can benefit from that, because maybe the
111 way, for example, I say something, someone else, might
112 not understand what I said, but if Carl repeats it in his
113 own words, then maybe someone else will understand the
114 same thing.

The instructor asked the class if anyone else wanted to add any ideas about how Ms. Hart supported students' learning of points of intersection. No one raised their hand, so the instructor moved on to the next idea, how Ms. Hart supported students' understanding of slope. Elaine was the first to speak, as shown in the following excerpt:

115 Elaine: Back on line 135[4], Danielle spoke up and said that she
116 had another observation. "She noticed they all had
117 different slants." And then Ms. Hart asked her to say more
118 about that, and Danielle really isn't quite sure what that
119 means, so [Ms. Hart] just kind of waits a little bit, and she
120 moves on and says, "we'll come back to that," and I think
121 that that's important as well, because at that particular
122 point, I don't think it was the place to go into that
123 discussion. She comes back to it when they bring that
124 back up, and she supports it by really linking back to
125 Danielle's first observation.
126 Instructor: So, this came up as an observation. [Ms. Hart] asked if
127 anybody had anything to say about it. They didn't. She
128 left it alone. Two hundred lines later, they come back and,
129 Carlos works out something. "It seems like the bigger the
130 number being multiplied by *N*, the steeper the graph." She
131 asks if this has anything to do with Danielle's idea, so she
132 ties it back, and then what? I mean, what I'm trying to get
133 at is, those are two important steps, in trying to move this
134 along, but if she just stopped at that point, would they
135 have really learned anything about slope, or laid this
136 important groundwork that Elaine's talking about.
137 Instructor: Melanie?

138 Melanie: When Danielle said, "the number multiplied by *N*, tells
139 you how steep it is," Ms. Hart wanted her to say more,
140 which brought out something specific to the problem.
141 Danielle's statement was kind of general, and she asked
142 her to say more, and then because she asked her to say
143 more, Danielle actually relates it to the problem and says
144 that "the number multiplied by *N* causes the total cost to
145 go up faster," so it gets more specific related to the
146 problem, so it has more meaning. So that was something
147 she did, rather than just dropping it in this general way.
148 Instructor: And it also seems interesting, that [Ms. Hart] just didn't
149 say, "say more", but she said, "say more about the
150 relationship", so it helped give Danielle some direction in
151 which to expand her explanation.

Two aspects of these excerpts from the case discussion are particularly noteworthy: 1) the ways in which teachers identify key accountable talk moves made by Ms. Hart; and 2) the ways in which the case discussion moves beyond the particulars of the case. During the case discussion, teachers identified pedagogical moves made by Ms. Hart that served to support her students' learning of mathematics – many of which could be considered accountable talk. For example, Bruce noted that Ms. Hart asks multiple students to contribute to a particular idea (lines 65-69). Later in the case discussion, Elaine argued that Ms. Hart's revisiting and linking back to Danielle's ideas later in the lesson was important (lines 115-125). Melanie also noted that Ms. Hart's press to Danielle to "say more" helped clarify her ideas about slope for the class (lines 141-147).

Second, during the case discussion, teachers moved beyond the particulars of *The Case of Edith Hart* to generalize pedagogical moves that support students' learning of mathematics. For example, Bruce noted similarities between the ways Ms. Hart and the instructor of the course facilitate discussions (lines 55-59; 61-63; 65). Recognizing an opportunity to make generalizations, the instructor asked, "You're identifying something that's really important, both in this lesson, and maybe bigger than that, pedagogically. So why did I do that?...Why does she do that, go around the room?" (lines 70-73). The teachers then considered how this move might support students' learning. For example, Lucy argued that such an approach would allow the teacher to build on students' current understandings, rather then telling students information (lines 76-79). Lisa noted another benefit of this approach, that when multiple students articulate an idea, others might understand one students' articulation better than another, thus more students have an opportunity to understand what is being discussed (lines 108-114).

In addition, teachers' comments during the case discussion indicate that they were thinking more generally about Ms. Hart's moves. For example, Melanie appeared to be considering why these moves might be important to her own teaching when she commented, "maybe she does it because if a student can articulate an idea on their own…then **you** really are sure they really understand it…So if she asks them to say what they mean, and then someone else repeats it, then **you** know they really understand it" (lines 81-87, emphasis added). Bruce also appeared to be making similar generalizations to his practice, when he noted, "I think **you** have to look out over **your** audience, and kind of see people's faces" (lines 98-99, emphasis added).

Teacher Learning from *The Case of Edith Hart*

As teachers engaged in activities centered on *The Case of Edith Hart*, they were held accountable by the instructor as they participated in a discussion of the mathematical task featured in the case and as they discussed the case. Through their reading and discussion of the case, teachers had opportunities to consider pedagogical moves (e.g., revoicing, pressing for clarification, holding all students accountable for understanding others' mathematical thinking) that can be critical in facilitating an effective mathematical discussion. We take as evidence of learning that teachers were able to identify such accountable talk moves in the case itself – and more importantly, that they recognized the instructor's use of these moves in the course and discussed how these moves can support students' learning of mathematics – both with respect to *The Case of Edith Hart* and their own practice.

Data collected at the end of the course provides further evidence that *The Case of Edith Hart* was an impetus in making salient the importance of accountable talk moves in facilitating effective mathematical discussions. For example, a discussion that occurred during the final class meeting, in which teachers were asked to consider the six episodes of teaching they had discussed during the course and identify the "lessons learned," provides additional evidence that teachers came to see the importance of accountable talk. During this discussion, Terry noted that her small group "mentioned the importance of having students justify their reasoning with mathematics." When asked by the instructor which episodes of teaching highlighted that idea, Terry identified *The Case of Edith Hart* as one episode that made this idea salient.

In addition, during an interview conducted at the end of the course, teachers were asked to reflect on their learning and describe what they felt they learned from participating in the course. Several teachers identified accountable talk moves as effective pedagogy when discussing what they learned about teaching mathematics. For example, Bruce noted:

> …asking another student to restate what another student had just said. Trying to get a few different opinions of the exact same statement and there are a few different ways of saying the same statement. So that the students can hang onto that, or they can grab onto whichever one makes sense to them.

In addition, as Kerry describes what she learned about teaching mathematics, she identifies accountable talk moves and makes explicit connections back to

The Case of Edith Hart:

> Well, I guess, questioning. Not giving away too much…questioning that gets out of the student what you want, but you're not putting an idea into their head...I keep going back to Edith Hart, but I like the way she taught…She was, "Well, what do you mean by that? Show me on the graph." You're not really giving them any answers, but I think it was an evaluation of what the student knew, and I think it was a verification that SHE understood what the student meant. And I think it helps for the student to discuss it, too. To explain what they mean, rather than just, here's the answer. "Well, what do you mean by that? Can you show me that? How did you get that?" Those types of questions I think are important.

Conclusion

At the beginning of this chapter we argued that accountable talk is useful for facilitating mathematical discussions that support students' learning. Furthermore, the rich dialogue often found in cases that depict classroom discussions can provide teachers with opportunities to identify accountable talk. In this chapter, we offer evidence that teachers' work on *The Case of Edith Hart* provided them with opportunities to develop understandings about accountable talk. The case discussion indicates that teachers were able to identify accountable talk moves and describe how such moves may impact students' learning. In addition, some teachers appeared to make connections to their own practice during the case discussion. Finally, when interviewed about what they had learned about teaching mathematics from the course, some teachers identified pedagogical moves that would be considered accountable talk moves (e.g., repeating or revoicing students' ideas; asking multiple students to express mathematical ideas in their own words; using questions that press students to clarify and justify their mathematical thinking).

Although the use of a case that featured a teacher holding her students accountable in order to support their learning was critical to provide teachers with opportunities to learn about accountable talk, the instructor's facilitation of teachers' work also seems critical to support teachers' work around a case. As teachers extended the mathematical ideas embedded in the case and participated in a discussion of the case, the instructor modeled effective pedagogy, and in particular, utilized accountable talk throughout the activities related to *The Case of Edith Hart*. For example, the instructor used revoicing during teachers' mathematical work (lines 22-25) to clarify teachers' statements. Revoicing was also used during the case discussion to highlight key elements of accountable talk (lines 70-73; 105-107; 126-132; 148-151). In addition, the instructor pressed teachers to justify their claims – both mathematical (lines 30; 32-33) and pedagogical (lines 130-136). Finally, the instructor capitalized on opportunities to move beyond the particulars of *The Case of Edith Hart* to generalities of pedagogical moves that support students' learning of mathematics (lines 70-75).

As noted at the beginning of this chapter, helping teachers consider the generalities that emerge from the analysis of a case is especially critical in developing teachers' understandings about teaching mathematics. In the case discussion described in this chapter, a teacher spontaneously noted the similarities between moves made by Ms. Hart and the instructor (lines 55-59; 61-63; 65), which led to a more general discussion of the moves, why one might make such moves, and how these moves might help students. But what if the teachers with whom you are working do not spontaneously make such comments? One possibility for helping teachers make generalizations would be to select a passage from the case in which you feel the case teacher uses accountable talk to support students' learning (or another pedagogical move that you want teachers to consider more carefully), and ask teachers what the case teacher was doing to support students' learning. You might then help teachers move beyond the particulars of the case by asking why this move was important or how this move might benefit students. Another promising approach might be to make your own pedagogy the object of inquiry (Steele, 2006). For example, you might ask teachers to consider what *you* were doing during a particular aspect of a lesson (e.g., What was I doing while you were working in small groups?; What was I doing during our whole class discussion?). By asking teachers to examine your practice, they have opportunities to consider how and why those moves serve to support their learning. In addition, they have opportunities to consider how the pedagogical moves you make, such as using accountable talk, might impact the learning of their own students.

References

Ball, D. L. (2001). Teaching, with respect to mathematics and students. In T. Wood, B. Nelson, & J. Warfield (Eds.), *Beyond classical pedagogy: Teaching elementary school mathematics* (pp. 11-22). Mahwah, NJ: Lawrence Erlbaum Associates.

Ball, D. L., & Cohen, D. K. (1999). Developing practice, developing practitioners: Towards a practice-based theory of professional education. In G. Sykes & L. Darling-Hammond (Eds.), *Teaching as the learning profession: Handbook of policy and practice* (pp. 3-32). San Francisco: Jossey-Bass.

Barnett, C. (1991). Building a case-based curriculum to enhance the pedagogical content knowledge of mathematics teachers. *Journal of Teacher Education*, *42*(4), 263-271.

Barnett, C. (1998). Mathematics teaching cases as a catalyst for informed strategic inquiry. *Teaching and Teacher Education*, *14*(1), 81-93.

Carpenter, T. P., & Lehrer, R. (1999). Teaching and learning mathematics with understanding. In E. Fennema & T. Romberg (Eds.), *Mathematics classrooms that promote understanding* (pp. 19-32). Mahwah, NJ: Lawrence Erlbaum.

Chazan, D., & Ball, D. L. (1999). Beyond being told not to tell. *For the Learning of Mathematics, 19* (2), 2-10.

Hiebert, J., Carpenter, T. P., Fennema, E., Fuson, K. C., Wearne, D., Murray, H., et al. (1997). *Making sense: Teaching and learning mathematics with understanding*. Portsmouth, NH: Heinemann.

Lampert, M. (2001). *Teaching problems and the problems of teaching*. New Haven, CT: Yale University Press.

Leinhardt, G., & Steele, M. (2005). Seeing the complexity of standing to the side: Instructional dialogues. *Cognition and Instruction, 23*(1), 87-163.

Merseth, K. K., & Lacey, C. A. (1993). Weaving stronger fabric: The pedagogical promise of hypermedia and case methods in teacher education. *Teaching & Teacher Education, 9*(3), 283-299.

Michaels, S., O'Connor, M. C., Hall, M. W., Resnick, L. B., & Fellows of the Institute for Learning (2002). *Accountable Talk$_{SM}$: Classroom conversation that works* [CD-ROM]. Pittsburgh, PA: Institute for Learning, Learning Research and Development Center, University of Pittsburgh.

National Council of Teachers of Mathematics. (2000). *Principles and standards for school mathematics*. Reston, VA: Author.

O'Connor, M. C., & Michaels, S. (1993). Aligning academic task and participation status through revoicing: Analysis of a classroom discourse strategy. *Anthropology and Education Quarterly, 24*(4), 318-335.

O'Connor, M. C., & Michaels, S. (1996). Shifting participant frameworks: Orchestrating thinking practices in group discussions. In D. Hicks (Ed.), *Discourse, learning, and schooling* (pp. 63-103). New York: Cambridge University Press.

Resnick, L. B. (1999, June 16). Making America smarter. *Education Week*, 38-40.

Shulman, L. S. (1996). Just in case: Reflections on learning from experience. In J. Colbert, K. Trimble, & P. Desberg (Eds.), *The case for education: Contemporary approaches for using case methods* (pp. 197-217). Boston, MA: Allyn & Bacon.

Smith, M. S. (2001). *Practice-based professional development for teachers of mathematics*. Reston, VA: National Council of Teachers of Mathematics.

Smith, M. S., Silver, E. A., & Stein, M. K. (with Henningsen, M. A., Boston, M., & Hughes, E. K.). (2005). *Improving instruction in algebra: Using cases to transform mathematics teaching and learning*. New York: Teachers College Press.

Steele, M. D. (2006). *Middle grades geometry and measurement: Examining change in knowledge needed for teaching through a practice-based teacher education experience.* Unpublished doctoral dissertation, University of Pittsburgh.

---

[1]This work was supported in part by a grant from the National Science Foundation (0101799) for the ASTEROID Project. Any opinions expressed herein are those of the authors and do not necessarily represent the views of the Foundation.

[2] The course was developed under the auspices of the ASTEROID project (directed by Margaret S. Smith), whose purpose was to design teacher education courses that made use of cases and other practice-based materials and to examine teachers' learning from participating in the courses.

[3] All teachers have been given pseudonyms.

[4]The paragraphs in *The Case of Edith Hart* are numbered so that teachers can efficiently cite evidence from the case to back up their claims.

Amy F. Hillen is an Assistant Professor of Mathematics at Robert Morris University. She earned her doctorate in mathematics education from the University of Pittsburgh in 2005. Her dissertation research focused on examining preservice secondary teachers' learning about proportionality as they participated in a course focused on proportional reasoning. She worked as a graduate student researcher on the NSF-funded COMET project, which developed middle grades cases and facilitation materials (including *The Case of Edith Hart)*, and the ASTEROID project, which examined teachers' learning from participating in teacher education courses that made use of the COMET cases and other practice-based materials. Her areas of interest include designing practice-based learning experiences for teachers and examining how such experiences influence their mathematical knowledge for teaching.

Elizabeth K. Hughes earned her doctorate in mathematics education from the University of Pittsburgh in 2006. Her dissertation research focused on lesson planning as a vehicle for improving preservice secondary teachers' attention to students' mathematical thinking. She worked as a graduate student researcher on the NSF-funded COMET project, which developed middle grades cases and facilitation materials (including *The Case of Edith Hart),* and the ASTEROID project, which examined teachers' learning from participating in teacher education courses that made use of the COMET cases and other practice-based materials. Her areas of interest include preservice secondary mathematics teacher education and the use of practice-based materials in developing teachers' understanding of what it means to teach and learn mathematics.

Silver, E. A., Clark, L. M., Gosen, D. L., and Mills, V.
AMTE Monograph 4
Cases in Mathematics Teacher Education: Tools for Developing Knowledge Needed for Teaching
©2008, pp. 89-102

8

Using Narrative Cases in Mathematics Teacher Professional Development: Strategic Selection and Facilitation Issues[1]

Edward A. Silver
University of Michigan

Lawrence M. Clark
University of Maryland – College Park

Dana L. Gosen
University of Michigan; Oakland (MI) Intermediate School District

Valerie Mills
Oakland (MI) Intermediate School District

One example of a practice-based approach to mathematics teacher professional development is the use of narrative cases of mathematics classroom lessons to stimulate inquiry into teaching and learning. Given that limited time is allotted for teacher professional development, strategic choices are critical regarding the selection of cases and the aspects of a chosen case on which to focus teachers' attention during facilitation. In this chapter, we draw on our experience in BIFOCAL (Beyond Implementation: Focus on Challenge and Learning) -- a multiyear, practice-based professional development project for teachers of mathematics in the middle grades -- to illustrate some issues and challenges encountered in selecting and facilitating narrative case. We highlight three general principles that we believe can apply across a variety of professional development settings to guide matters of case selection and facilitation.

Since the mid-1990s, teacher educators, professional developers, and researchers have taken great interest in the development and facilitation of *practice-based* approaches to mathematics teacher education (Smith, 2001). In practice-based learning environments, teachers engage with tasks that embody authentic aspects of instructional practice allowing teachers to access, utilize and develop knowledge of mathematics content, pedagogy, and student learning *simultaneously* (Ball & Cohen, 1999). In these environments, teachers study and engage in a range of activities deliberately situated in 'the work of teaching' – activities that resemble or replicate components of teachers' daily work, such as planning for mathematics instruction, analyzing student work, and viewing and discussing instructional episodes. In short, practice-based approaches to mathematics teacher education promote development of the knowledge teachers need by engaging them with tasks and situations that embody the complex interactions that occur in their classrooms (Smith, 2001).

In recent years, mathematics teacher educators have developed and utilized a number of approaches to practice-based professional development. The approach of interest here is the use of narrative cases of mathematics instruction as the basis for professional learning tasks. Narrative instructional cases, particularly those developed in the QUASAR and COMET projects (Smith, Silver & Stein, 2005a, 2005b, 2005c; Stein, Smith, Henningsen, & Silver, 2000), offer accounts of mathematics instructional episodes depicting interactions that occur when a teacher uses a complex mathematical problem in the classroom, with attention to the teacher's actions and interactions with students as they solve the problem. Cases of this form are deliberately constructed to provoke discussion regarding interactions among the teacher, the students, and the mathematical task, and the way in which those interactions affect students' opportunities to learn mathematics. In addition, the cases provide opportunities for teachers to consider, and sometimes learn, mathematics content as they complete an opening activity in which they solve the mathematics task featured in the narrative case, participate in an analytic discussion of their solution(s), and read and discuss the narrative case to see how the mathematics appears in the work of the teacher and students in the lesson depicted in the case. As a result, mathematics instructional cases in the form of well-crafted narratives have potential as the basis for professional learning tasks in practice-based professional development endeavors focused on enhancing knowledge of mathematics, pedagogy, and students.

Why Focus on Case Selection and Facilitation?

Despite the evident potential of narrative cases as powerful professional learning tasks, little has been written about issues that arise regarding the selection and facilitation of cases in particular professional development contexts. Given that limited time is allotted for teacher professional development, strategic choices are critical about the cases to choose and the aspects of a chosen case or set of cases to emphasize during facilitation. Naturally, these choices should be driven by the goals of the professional development. But many professional development initiatives are complex endeavors, and each session will likely require a mix of attention to short-term goals (immediate desired outcomes related to teachers' engagement at a single session) and long-term goals (desired outcomes related to the cumulative experience of teachers' engagement over time). Although short-term and long-term goals may align neatly when they appear on paper, there may well be tensions between them during the enactment of professional development. For example, a facilitator who is trying to foster broad participation among a group of teachers may tolerate some "off the mark" discussion in a session in the service of that goal, even though this might, for example, compromise progress toward a different goal (e.g., having teachers argue from evidence rather than opinion).

In many ways the issues, challenges, and dilemmas faced by facilitators of professional development regarding the selection and facilitation of cases parallel those faced by teachers using complex mathematics tasks in the classroom. Like teachers, professional developers must select a worthwhile task, present it in a way that engages participants in thoughtful activity, resist the urge to "move things along" by doing most of the thinking and talking rather than allowing participants to do so, and assist participants as they integrate and synthesize new ideas. As is the case for teachers and children in discourse-intensive mathematics classrooms (e.g., Silver & Smith, 1996), the facilitator must be attentive to issues of time, focus, and coherence within a single session, as well as integration and flow across multiple sessions. Thus, facilitators must make frequent decisions based on their calibration and recalibration of the balance among their goals for a particular professional development session, their long-term goals for the professional development initiative, and their desire to respond to teacher needs and interests.

In what follows we illustrate some of the issues and challenges faced in selecting and facilitating narrative cases in mathematics professional development, drawing on our experience in BIFOCAL (Beyond Implementation: Focus on Challenge and Learning), a multiyear, practice-based professional development project for teachers of mathematics in the middle grades. We do so because we believe that our BIFOCAL experience can inform users of cases under varied circumstances. We use excerpts from our work in BIFOCAL to illustrate three general principles we believe might apply across a variety of professional development settings in which cases are used:

1. Facilitators should choose cases that are likely to direct teachers' attention toward worthwhile mathematical tasks and toward teachers' classroom actions and interactions that influence students' opportunities to learn mathematics through engagement with the tasks.
2. Facilitators should orchestrate case discussions in a manner that encourages participants to focus their attention on important mathematical ideas and associated pedagogical opportunities, issues, and challenges.
3. Facilitators should listen and respond to participants' emergent needs, concerns, and interests with an eye toward deconstructing, negotiating, and reconstructing goals as needed to adapt the flow of a session or the flow across sessions to accommodate emergent themes.

The BIFOCAL Project

BIFOCAL was designed to support teachers who were users of standards-based middle school mathematics curriculum materials.[2] These materials provide a rich source of worthwhile, complex mathematics tasks for use in instruction, but teachers often face challenges in using such tasks effectively to provide powerful learning opportunities for their students. BIFOCAL built on a foundation of prior work in the QUASAR and COMET projects, particularly research regarding the use of cognitively demanding tasks in the classroom (e.g., Henningsen & Stein, 1997; Stein, Grover, & Henningsen, 1996) and narrative instructional cases designed to draw attention to issues and challenges faced by teachers in using such tasks (e.g., Stein, Smith, Henningsen & Silver, 2000). Each BIFOCAL session engaged participants in two interrelated activities: (1) case analysis and discussion and (2) modified lesson study.[3]

Because of our focus on the effective use of cognitively challenging tasks, the Mathematical Task Framework (MTF) (Stein, Grover & Henningsen, 1996), was used as a core framework in BIFOCAL.[4] The MTF (see Figure 1) portrays the passage of instructional tasks as they move from the pages of curriculum materials into the work of teachers and students in classrooms, with special attention to the modifications that teachers may make, intentionally or unintentionally, to reduce the cognitive complexity of challenging tasks. The MTF thus draws attention to the importance of a teacher's actions and interactions in supporting students' engagement with cognitively demanding tasks. In addition, the MTF points to the critical importance of both the cognitive demand of tasks and the role teachers play in influencing the opportunities that students have to learn mathematics through engagement with complex tasks.

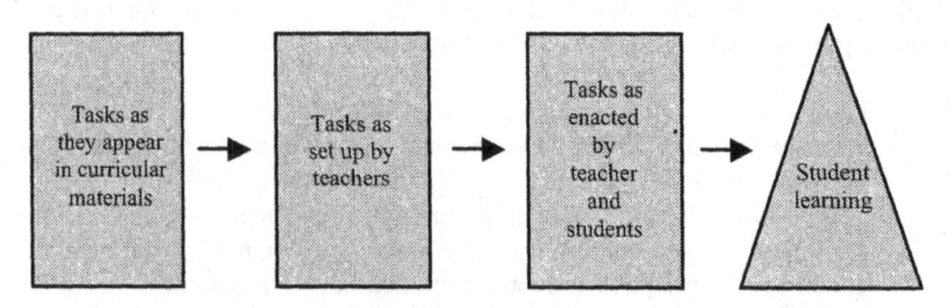

Figure 1. Mathematical Tasks Framework (MTF)

Selecting and Facilitating Cases in BIFOCAL

To illustrate the interplay between professional development goals and the selection and facilitation of cases, we focus on three professional development sessions during the first year of the project: May 2003, June 2003, and February 2004.

May 2003 Session

Our short-term goal for this inaugural BIFOCAL session was to offer participants an opportunity to consider the core issues of the MTF as they pertain to actual mathematics lessons. We decided to use *Examining Linear Growth Patterns: The Case of Catherine Evans and David Young* (Smith et al., 2005a, pp. 10- 27) as the first case for the participants to read and discuss. We did so because our experience with this case in prior professional development demonstrated its power to stimulate productive discussion among teachers about a range of instructional issues connected to the MTF.

This case depicts two teachers (Catherine and David) who approach the enactment and facilitation of a cognitively demanding mathematics task in distinctly different ways. The contrasting instructional approaches draw a reader's attention to pedagogical issues and challenges that arise when complex mathematical tasks are used in the classroom. In addition, the case affords an opportunity to inquire about the mathematical goals of each teacher and the ways in which each teacher's classroom actions and reactions during task enactment relate to students' opportunities to learn mathematics. In particular, Catherine intends that her students experience success in her classroom, and this leads her to structure the problem-solving experience of her students in ways that help them avoid struggling. In so doing, she reduces the cognitive demand of tasks in order to enable students to be successful. In contrast, David does not lower the cognitive demand. Instead, he gives his students time to think and to make sense of a complex task, and he supports them in ways that do not reduce the challenge of the task. The participants examined these two different approaches to facilitating a cognitively demanding mathematics task and considered how each approach inhibited or enhanced students' opportunities to learn mathematics. The sharp contrast between the two teachers depicted in the case, whose teaching also shares many similarities, makes this an accessible and productive first experience with narrative case analysis and discussion.

Before reading the case, participants began by solving a mathematics task that is central to the case: finding the perimeter of a "pattern block train" formed by x adjacent hexagonal "cars" where $x = 1, 2, 3, 4, 10$, and n "cars" (see the Appendix for details). After having time to work individually, participants presented different solution strategies to the group. Participants also thought about how their students would approach and solve this task. Exploration of the mathematics task outside of an instructional context was intended to engage the participants in some mathematical work and to set the stage for their reading, analysis, and discussion of the case.

After considering the mathematics embedded in the case through this initial problem solving, participants read the narrative case and considered the following questions: "In what ways were Catherine's and David's instructional choices and actions similar and different? Do the differences make a difference in students' mathematical learning opportunities? Why or why not?" In general, the facilitated case discussion touched on the desired points, drawing attention to the similarities and differences in the two teachers' instructional moves and examining the case for evidence of any consequential impact of these differences.

The session served well its goal of stimulating attention to issues related to the MTF and surfacing instructional issues that resonated with participants. One of these issues was the instructional practice of having students present "multiple solution strategies" as a challenge of some concern and disagreement among participants. Another was the difficulty that participants had in parsing the mathematical content of a complex task, especially with respect to the overarching ideas that could frame its use in a lesson or unit of instruction. Thus, not only were the mathematics and case discussions engaging for teachers, they also served as an important formative assessment for the facilitators. We provide the following glimpse at the May 2003 session to give the reader samples of the conversation regarding multiple solution strategies and of the thinking of the facilitators while orchestrating the discussions.

During the initial problem solving, the participants produced several different solution methods and publicly mentioned various anticipated student approaches. The discussion of multiple solution strategies prompted one teacher to ask a question about the value of considering multiple solutions in her classroom: "Do all students benefit from considering more than one solution?" After a brief exchange of differing views, it was evident that most teachers supported the classroom activity of students sharing multiple solution approaches, though the basis for support seemed unclear beyond "validating the thinking of everyone" and valuing "different ways of thinking." Although this issue appeared to strike a responsive chord among the participants, the facilitator decided not to provoke a deeper discussion at this time. His decision was influenced by time constraints, given that more time had been spent on this segment of activity than was originally planned, and by a concern that dwelling on this particular issue might detract from the more general discussion of MTF issues in relation to the case. Also, the facilitator was confident that the issue of multiple solutions would surface again during the analysis and discussion of the narrative case.

When reading the case of Catherine and David, participants encountered an instructional issue that had considerable resonance: How can a teacher help students succeed when they are working on a challenging task? The apparent familiarity of this core instructional issue appeared not only to support participants in examining the instruction depicted in the case and analyzing the similarities and differences between the two instructional episodes but also to support the participants in considering this issue in relation to their own mathematics teaching. The case discussion in May 2003 created many opportunities for participants to move between the "fictional" space of the case and the realities of their own classroom practice.

The issue of multiple solution strategies arose again quite naturally in relation to ways that Catherine and David did or did not invite consideration of different student solutions in their lessons. Teachers expressed support for teaching moves in the case that supported students' engagement with multiple solution strategies. However, when the teachers were pressed to explain how and why students' consideration of multiple strategies might affect their opportunities to learn, they had relatively few justifications. One participant commented: "...instead of focusing on just one student, you help everyone feel comfortable to give their opinion, or share their strategy or their way of how they looked at it..." This comment exemplifies a fairly common view across the group, namely, multiple solution approaches are desirable for affective reasons but are not necessarily tied to the learning of mathematical content. Further discussion of this issue allowed some concerns to surface. Some participants voiced concerns that "below average" students might be confused by multiple solution paths; others voiced concerns about the time needed for this practice.

Reflecting on May and Preparing for June

Our reflection on the May 2003 session revealed several issues that appeared to warrant further attention in subsequent sessions. For example, handling multiple solution strategies in the classroom clearly emerged as an instructional issue that participants perceived as important but problematic. We judged that spending time in subsequent sessions on the effective use of multiple solutions would be wise for two reasons: (1) it might support teacher investment in BIFOCAL given that this issue was raised by the participants; and (2) this aspect of teaching is important for scaffolding student learning with and through complex mathematics tasks.

Although we thought that further work on multiple solution strategies would advance the long-term goals of the project, we were less certain about when and how to revisit this issue. If this were the only issue that had emerged as needing attention in our reflection on the May session, life would have been simple. Alas, this was not the case. A number of other issues needed attention, and so we had to prioritize and negotiate subsequent session content. An additional concern was our perception that participants were not facile in identifying an overarching or "big" mathematical idea associated with a complex task. In the May session, participants tended to make superficial, topical associations with tasks rather than identifying deep, important mathematical ideas. Moreover, it was clear that more work would be needed to amplify participants' understanding of the MTF.

After considering various options, we decided to make explicit in our next session the matter of associating tasks and lessons with important mathematical ideas. We judged that discussions concerning the MTF and associated instructional moves, such as the effective use of multiple solution strategies, would be impeded by confusion about clearly identified mathematical goals. We viewed the MTF as depending to some extent on the idea that a teacher's instructional decisions regarding task selection, presentation, and enactment should be guided by the teacher's mathematical goals. Thus, we decided to defer further explicit treatment of the use of multiple solutions in the classroom, although we certainly intended that the issue would be treated if it arose in connection with the initial problem solving or case discussion in the next session.

June 2003 Session

The two-day session in June 2003 featured the analysis and discussion of two cases, an opportunity for modified lesson study, and general discussions related to aspects of maintaining cognitively demanding tasks in the mathematics classroom. The short-term goals associated with the June 2003 session built on the goal of the previous session – increasing participants' familiarity with and use of the structure and language of the MTF – and incorporated a focus on identifying the mathematical goals the case teachers implicitly or explicitly pursued. We selected two cases to provide an opportunity for participants to work on these issues.

The first case was *Linking Fractions, Decimals, and Percents Using an Area Model: The Case of Ron Castleman* (Stein et al., 2000, pp. 47-56). The case portrays a seventh-grade teacher (Ron) as he facilitates students' engagement with a task that requires them to use visual representations to determine the percent, fraction, and decimal corresponding to a portion of a shaded area of a 4-by-10 rectangular grid. The case depicts Ron's attempts to support his students' work without telling them precisely how to solve the problem, thereby keeping the enactment of the task at a high level of cognitive demand. In addition, this case offers readers a dichotomous situation similar to that of Catherine and David in that it depicts Ron using the same task twice, with different groups of students. The dual enactments sharpen a reader's attention to the MTF, by making visible the relationship between a teacher's instructional decisions and students' opportunities to learn. Facilitators can use "dual cases" to press even MTF novices to make relevant observations that can be supported by evidence in the narrative.

We prepared focus questions that would directly link the MTF to the Ron Castleman case and bring explicit attention to the matter of mathematical goals. Participants were asked to consider each of the following focus questions as they reflected on the case.

1. Why do you think that Ron's task asks for percent, followed by decimal, and finally fraction? How does this ordering relate to his mathematical goal(s) for the lesson? Would a different ordering be likely to make a difference? If so, how?
2. Identify specific examples of teaching decisions from the Ron Castleman case that:
   a. Appeared to help *maintain* the cognitive demand of the task and/or
   b. Appeared to *undermine* the cognitive demand of the task.
3. In what phase of the MTF did each of the teacher's moves you identified in (2) occur?
4. If time permits, for each of the examples you identified in (2) look for specific evidence of the impact the teacher's move may have had on student learning.

The facilitator used the focus questions to direct participants' attention to the mathematics, and more specifically the nature and depth of the mathematical ideas that can be explored as a result of the task and the teacher's moves in each of the two lessons depicted in the case. The focus questions also helped the facilitator remain true to the specific intent of the session. By the end of the discussion of the Ron Castleman case, it was apparent that participants were becoming more comfortable with the language of the MTF and its phases, and thus, better equipped to recognize the mathematical goals explicitly or implicitly held by the teacher in the case. For example, they discussed the linearity of the MTF phases as depicted in the diagram. Participants commented that they felt Ron's teaching challenged a simple linear view of the process. He set up the problem as it appeared in the curriculum materials and only after considerable time enacting the task with students did he restructure it. As a result, teachers began to grapple with such questions as: When does a phase start and when does it end? If a teacher adds a piece of information to the task after students start working with it, does this mean that he is re-setting the task? What is the teacher's role in the enactment phase? How do instructional decisions affect the students' work and their learning opportunities?

The second case, *Multiplying Fractions with Pattern Blocks: The Case of Fran Gorman and Kevin Cooper* (Stein et al., 2000), offered participants additional opportunities to explore the mathematical goals embedded within the case, the nature of the MTF, and a teacher's role in maintaining or undermining the cognitive demand of mathematical tasks. Our focus questions were similar to those for the Castleman case.

February 2004 Session

Between June 2003 and January 2004, we continued to work with participants on the identification of mathematical goals and the coordination of instructional decisions with those goals. All of this was done within the framework provided by the MTF. Some narrative cases, one video case, and one structured collection of student work samples provided the basis for the participants' mathematical problem solving and their analysis and discussion of teaching.

In planning for the February 2004 session, we determined that it was time to revisit the issue of multiple solution strategies. A discussion analyzing the effectiveness of teaching with multiple solution strategies initially surfaced in the May 2003 session and reappeared during the January 2004 session (not discussed here). Because this issue appeared to be resilient and because the participants were now better able to identify important mathematical goals associated with instructional tasks, we decided to select a case in which the use of multiple solution strategies would emerge as a focal point for analysis and discussion. At this point we hoped to link the issue of multiple solutions more explicitly to the MTF with which participants were now familiar and comfortable. In particular, we hoped to underscore the importance of anticipating the many ways that students might approach a problem as an integral part of planning for instruction so a teacher would be better prepared to use solution approaches when they appeared in student work during a lesson.[5] In February we also wanted to focus on mathematical ideas associated with ratio and proportion. In prior sessions, especially in January 2004, we noted that participants' knowledge of proportionality seemed fragile and disconnected.

To accomplish both short-term goals for the February session, we selected *Introducing Ratios and Proportions: The Case of Marie Hanson* (Smith et al., 2005b, pp. 26-43). In this case, Marie Hanson, a sixth-grade teacher, uses a mathematics task that introduces her students to ideas of ratio and proportionality in the context of a complex problem. The students in Marie's class use a number of different strategies to solve the problem, including some that are not correct. The following focus questions reflect our attempt to maintain continuity with prior sessions and to bring into focus the issue of multiple solution approaches (including errors):

1. What are the teacher's mathematical goals for the students?
2. What kinds of student reasoning/solutions might we anticipate on this task?
3. What student misconceptions or errors might we anticipate with this task and what did she do to assess possible errors?
4. Speculate on Marie's rationale for the way in which she worked with the various student approaches and consider how those choices may have supported or undermined her mathematical goals.

In discussing the case, many participants commented on the moves Marie made in managing the multiple student solution approaches in her classroom. Participants became especially animated in grappling with the issue of incorrect solution approaches. In the narrative case Marie indicates that she is concerned about the "addition fallacy" students often make when generating equivalent ratios – adding the same constant to both quantities in the ratio to create an equivalent ratio rather than multiplying both quantities by a constant, or scale factor. This issue resonated with participants, and they debated about the relative merits of "bringing out the error" early in the presentation of solutions or near the end. The discussion of how to handle incorrect solution approaches had not occurred in prior considerations of this general issue in May 2003 and January 2004. Thus, the Marie Hanson case introduced a new element into the participants' discussion of this issue and provoked them to consider how errors might be used to leverage progress toward the mathematical goals for a lesson and a deeper understanding of the ideas associated with a complex task. In this way, discussion of the management of multiple solutions in the classroom was expanded and became even more nuanced.

Conclusion

The selection and facilitation of narrative mathematics cases is a complex undertaking that can be consequential for the success of a professional development endeavor. The three principles articulated at the outset - choosing cases that draw attention to worthwhile mathematical tasks, focusing participants' attention on important mathematical ideas, and adapting the flow of a session or the flow across sessions to accommodate emergent themes - can help facilitators navigate this complexity. Thoughtful facilitation is of critical importance when engaging teachers in cases because, like all professional learning tasks in practice-based professional development, cases are not "self-enacting" and can serve multiple purposes depending on the interests of the teachers and goals of the facilitator.

We were fortunate because BIFOCAL consisted of multiple sessions across a year, each consisting of full-day sessions that allowed considerable time for analysis and discussion of cases and other practice-based activities. These features allowed for flexibility in our approaches and responsiveness to teacher needs. Although not all professional development endeavors have this long-term character, or perhaps *because* they do not have it, we think that purposeful selection and skillful facilitation of cases are critically important whenever cases are used in professional development.

References

Ball, D. L., & Cohen, D. (1999). Developing practice, developing practitioners. In L. Darling-Hammond & G. Sykes (Eds.), *Teaching as the learning profession* (pp. 3-32). San Francisco, CA: Jossey-Bass Publishers.

Henningsen, M. A., & Stein, M. K. (1997). Mathematical tasks and student cognition: Classroom-based factors that support and inhibit high-level mathematical thinking and reasoning. *Journal for Research in Mathematics Education, 28*, 524-549.

Hughes, E. K., & Smith, M. S. (2004). *Thinking through a lesson: Lesson planning as evidence of and a vehicle for teacher learning*. Paper presented at the annual meeting of the American Educational Research Association. San Diego, CA.

Lappan, G., Fey, J. T., Fitzgerald, W. M., Friel, S. N., & Phillips, E. D. (1996). *Comparing and scaling: Connected Mathematics Project*. Palo Alto, CA: Seymour.

National Council of Teachers of Mathematics. (2000). *Principles and standards for school mathematics*. Reston, VA: Author.

Silver, E. A., Mills, V., Castro, A., & Ghousseini, H. (2006). Blending elements of lesson study with case analysis and discussion: A promising professional development synergy. In K. Lynch-Davis & R. L. Rider (Eds.), *The work of mathematics teacher educators: Continuing the conversation* (AMTE Monograph Series, Volume 3) (pp. 117-132). San Diego, CA: Association of Mathematics Teacher Educators.

Silver, E. A., & Smith, M. S. (1996). Building discourse communities in mathematics classrooms: A challenging but worthwhile journey. In P. C. Elliott & M. J. Kenney (Eds.), *Communication in mathematics, K-12 and beyond* (pp. 20-29). Reston, VA: National Council of Teachers of Mathematics.

Smith, M. S. (2001). *Practice-based professional development for teachers of mathematics*. Reston, VA: National Council of Teachers of Mathematics.

Smith, M. S., Bill, V., & Hughes, E. K. (in press). Thinking through a lesson: The key to successfully implementing high level tasks. *Mathematics Teaching in the Middle School.*

Smith, M. S., Silver, E. A., & Stein, M. K. (2005a). *Improving instruction in algebra: Using cases to transform mathematics teaching and learning, Volume 2.* New York: Teachers College Press.

Smith, M. S., Silver, E. A., & Stein, M. K. (2005b). *Improving instruction in rational numbers and proportionality: Using cases to transform mathematics teaching and learning, Volume 1.* New York: Teachers College Press.

Smith, M. S., Silver, E. A., & Stein, M. K. (2005c). *Improving instruction in geometry and measurement: Using cases to transform mathematics teaching and learning, Volume 3.* New York: Teachers College Press.

Stein, M. K., Grover, B. W., & Henningsen, M. A. (1996). Building student capacity for mathematical thinking and reasoning: An analysis of mathematical tasks used in reform classrooms. *American Educational Research Journal, 33*, 455-488.
Stein, M. K., Smith, M., Henningsen, M. A., & Silver, E. (2000). *Implementing standards-based mathematics instruction: A casebook for professional development*. New York: Teachers College Press.

---

[1]This article is based upon work supported in part by the National Science Foundation under Grant No. 0119790 to the Center for Proficiency in Teaching Mathematics and in part by the Michigan State University Mathematics Education Endowment Fund for support of the BIFOCAL project. Any opinions, findings, and conclusions or recommendations, expressed in this material are those of the authors and do not necessarily reflect the views of the National Science Foundation, the Center, or the university. The authors are grateful to Hala Ghousseini, Alison Castro, Gabriel Stylianides, Kathy Morris, Beatriz Strawhun, Charalambos Charalambous, Jenny Sealy and Angela Hsu, who assisted with the planning of professional development sessions and/or in the collection and/or analysis of data used herein. We also wish to acknowledge the positive influence on our thinking provided by Geraldine Devine. Finally, we thank the teachers who participated in the BIFOCAL project and openly shared their teaching practice and their wisdom. Nevertheless, we assume responsibility for any errors of fact or interpretation made in this paper.
[2]BIFOCAL participants used *Connected Mathematics* (Lappan, Fey, Fitzgerald, Friel, & Phillips, 1996).
[3]Because of space limitations in this chapter, we do not provide details about the ways in which case analysis and lesson study were integrated in the project; more details may be found in Silver et al. (2006).
[4]For more on the MTF, see Stein et al. (2000).
[5]To support the planning of lessons in ways tied to the MTF framework, project participants adapted the Thinking Through A Lesson Protocol developed by Smith and colleagues (Smith, Bill, & Hughes, in press).

Edward Silver is the William A. Brownell Collegiate Professor of Education and the Associate Dean for Academic Affairs in the School of Education at the University of Michigan. He teaches and advises graduate students in mathematics education, conducts research on the teaching and learning of mathematics, and engages in a variety of professional service activities. He is co-PI of the NSF-funded Center for Proficiency in Teaching Mathematics and co-director of the BIFOCAL Project and the Michigan Mathematics and Science Teacher Leadership Collaborative, both of which support the learning of mathematics teachers in Michigan school districts.

Lawrence Clark is an Assistant Professor of Mathematics Education at the University of Maryland - College Park. He is co-PI of *Ongoing Professional Development in Mathematics*, an NSF-funded, online professional development project. Previously he was a mathematics teacher in the middle grades and a postdoctoral research fellow at the University of Michigan, where he served on the BIFOCAL Project team. He has also been National Director of Mathematics for Project GRAD USA, a national school reform initiative in several urban districts. His research focuses on mathematics instructional practice and on professional development for middle grades mathematics teachers and their administrators.

Dana L. Gosen is a Mathematics Consultant for Oakland Schools, a Michigan public schools resource agency serving 28 districts and 260,000 students. She is also a graduate student at the University of Michigan pursuing a doctoral degree in teacher education. In her work at the university she has served as a supervisor of student teachers, an apprentice to the elementary mathematics methods course, and a graduate student researcher on the BIFOCAL project. At Oakland Schools she provides mathematics support in the areas of curriculum, assessment, and instruction.

Valerie L. Mills is Mathematics Consultant and Learning Sciences Supervisor for Oakland Schools, a Michigan public schools resource agency serving 28 districts and 260,000 students. She supports mathematics programs countywide, providing resources, consulting, and professional development. She is currently co-director of the BIFOCAL project, a university–school professional development research collaborative, PI of the Mathematics Education Resource Center project serving under-performing urban schools, and a member of the Lenses on Learning: Secondary writing team.

Appendix

Hexagon-Pattern Task

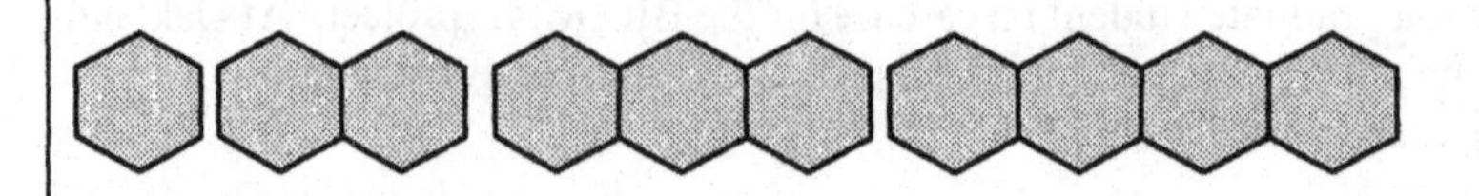

The first four pattern block "trains" in a sequence are shown above.
The first train in this pattern consists of one regular hexagon (car).
For each subsequent train, one additional hexagon (car) is added.
You should assume that the sequence continues indefinitely in this manner.

a) Using the edge length of a pattern block as your unit of measure, find the perimeter for each of the first four trains.
b) Without constructing the tenth train, determine its perimeter.
c) Give a description or formula that could be used to determine the perimeter of any train in the pattern.

Illustration from *Visual Mathematics Course 1, Lessons 16-30* published by The Math Learning Center, www.mathlearningcenter.org.

Romagnano, L., Evans, B., and Gilmore, D.
AMTE Monograph 4
Cases in Mathematics Teacher Education: Tools for Developing Knowledge Needed for Teaching
©2008, pp. 103-115

9

Using Video Cases to Engage Prospective Secondary Mathematics Teachers in Lesson Analysis[1]

Lew Romagnano
Brooke Evans
Don Gilmore
The Metropolitan State College of Denver

How can methods courses help prospective mathematics teachers make sense of the enormously complex work of teaching mathematics? We propose that supporting the ability of prospective teachers to engage in lesson analysis is an important and reasonable goal for these courses. Lesson analysis can help novices to see, hear, describe, discuss, and understand the interactions among teacher and students that comprise a lesson. By focusing on individual lessons as the units of analysis, we are able to address in considerable depth many aspects of teaching practice, including content and curriculum, learning, teaching, assessment, and context. We argue that lesson analysis supports prospective teachers' understanding of the relationships between teacher actions and their consequences for student learning, and builds a foundation for learning how to plan effective lessons. In this chapter we describe a preservice program in which prospective mathematics teachers analyze lessons they encounter in several settings: as students in mathematics lessons we teach; as observers and participants in field experience lessons; and as analysts of video and written cases. We use a video case as an example to illustrate the conceptual lenses we have chosen and how we use them with our students to help them make sense of a lesson.

There is little consensus among mathematics teacher educators about the goals, structure, or specific activities and assignments that comprise the preservice mathematics "methods" courses many of us, nonetheless, must teach.[2] There is, after all, so much to do to prepare mathematics teachers for the rigors of full-time teaching, such as covering a curriculum, planning and enacting lessons, assessing student learning and managing student behaviors. These are the same work demands experienced teachers face. How can our methods courses help beginning mathematics teachers make sense of the enormously complex work of teaching mathematics?

One common approach is to build such courses with the goal of providing prospective teachers with the skills of a competent beginning teacher. We think this goal is unrealistic. We agree with Hiebert and colleagues (Hiebert, Morris & Glass, 2003), who assert that "[P]reparation programs can be more effective by focusing on helping students acquire the tools they will need to learn to teach rather than the finished competencies of effective teaching" (Hiebert, Morris & Glass, 2003, p. 202).

We propose that supporting the ability of prospective teachers to engage in *lesson analysis* is a critical—and attainable—goal of preservice mathematics methods courses. Lessons are entities with a mathematical narrative arc and an internal structure; yet, for many prospective teachers, this structure is lost amidst the seeming chaos of events that constitute teaching practice. Lesson analysis, using a set of conceptual lenses and introducing a common language, can help novices to see, hear, describe, discuss and understand the interactions among teacher and students that comprise a lesson.[3]

The lesson is the conceptual unit prospective teachers can use to bring order to the barrage of information they confront when observing or participating in classroom instruction. It is a large enough unit of analysis to bring together all of the complexity of the classroom, yet it is a small enough unit to allow for in-depth study (Hiebert, Morris & Glass, 2003). Through comprehensive analysis of lessons, prospective teachers can:

- delve into important mathematical ideas;
- learn about mathematics curriculum, materials and tools;
- identify teacher actions and their consequences;
- assess student learning;
- assess the effectiveness of the lesson.

For prospective teachers who have no teaching experience, the understanding of lesson flow and structure, and of teacher actions and their influences on students that comes from lesson analysis provides an important basis for learning how to prepare and enact effective lessons, which is at the heart of the daily work of teaching. In the sections that follow, we use a particular example to illustrate how we use video cases as important components of our teaching and learning program to help our preservice teachers learn to do lesson analysis.

Context

Lesson analysis is an important component of our course called *Teaching Secondary Mathematics*. The course, one in a sequence of courses offered by our Department of Mathematical Sciences for mathematics majors seeking a secondary teaching credential, has both classroom and field-experience components. The undergraduate and post-baccalaureate prospective teachers in the class have completed a prerequisite course called *The Mathematics of the Secondary Curriculum*, in which they engage in *content analysis*; that is, they explore and analyze problems from the secondary mathematics curriculum from their advanced perspectives as mathematics majors near the end of their undergraduate programs. They will follow the *Teaching Secondary Mathematics* course with a semester-long, full-time student-teaching experience.

Participants in *Teaching Secondary Mathematics* encounter lessons in several settings: as learners in mathematics lessons we teach in the context of the course; as observers and participants in field experience lessons; and—the focus of this chapter—as analysts of video and written cases. Such cases play an important role in learning to do lesson analysis because, unlike the other "real time" lesson settings, video and written records of teaching practice (Smith, 2001) allow us to view the lesson repeatedly, posing questions and then revisiting specific important moments in the lesson in search of answers.

The *Teaching Secondary Mathematics* course begins with three lesson analysis activities, focused on lessons in three different settings—in our classroom (which we call "lessons within the lesson"), on videotape, and in the field. In each of these settings we employ a set of conceptual lenses through which preservice teachers view the lessons. We identify common features of the lessons and introduce specific language to discuss these features. Our preservice teachers analyze mathematical content, review student work, and begin to learn how to prepare and enact lessons. Following is a description of these lenses and why we chose them, and a discussion of a video case-based example to illustrate how we engage preservice teachers in lesson analysis.

Four Lenses for Viewing Lessons

We use four different lenses for viewing and understanding lessons. Each affords preservice teachers a particular perspective on the lesson, and taken together they give them some insight into their complexity.

The Temporal Lens

This is the most descriptive of the four lenses we use, and it is the way observers typically see lessons. We ask preservice teachers to identify the "sequence of events" in the lesson. How did the lesson begin? How did it end? Were there different segments, or "chunks," of the lesson? Many lessons, they learn, consist of four chunks: a warm-up, homework review, teacher exposition, and seatwork. But other possible chunks might include a period of time in which students work on a task in small groups, followed by some time devoted to whole-class discussion of students' solutions, strategies and difficulties. Preservice teachers use this lens to identify these chunks and their purposes, discuss the time spent and the student engagement during each chunk, and note the teacher's actions during each chunk and in the transition to the next one.

Lesson Dimensions

Following Artzt and Armour-Thomas (2001), we use a framework that identifies three dimensions of every lesson:

- Mathematical Tasks (representation, motivation, sequencing and difficulty)
- The Learning Environment (social norms and expectations, *sociomathematical* norms, instructional structures, strategies, administrative routines)
- Interactions (student – student, teacher – student, questioning)

The first dimension allows us to keep the mathematics in each lesson at the forefront, which is stressed even more with the next lens. Our discussions of the learning environment address the questions, "What does doing mathematics look like in this class?" and "From the perspective of the students in the class, what does it mean to do mathematics?" The third dimension highlights our commitment to the situative perspective that learning is a social phenomenon and that student-student and student-teacher interactions provide important insights into learning in classrooms (Cobb & Bowers, 1999).

***The Thinking Through A Lesson Protocol* (TTLP)**

We focus more deeply on the mathematics in a lesson, and how that mathematics is manifested in the tasks posed for students, using the Thinking Through A Lesson Protocol developed by Smith and colleagues at the University of Pittsburgh (Smith, Bill, & Hughes, in press). This lens places preservice teachers firmly in the role of the teacher who uses rich mathematical tasks to drive instruction, and asks them to consider these important aspects of that work:

- Selecting and Setting up a Mathematical Task: what are the mathematical goals; what is the important content; what prior knowledge will students draw on; what are some possible solutions and strategies; what are my expectations for students?
- Supporting Students' Exploration of the Task: what focusing, assessing, and advancing questions can I ask to support students' work?
- Sharing and Discussing the Task: how will I orchestrate the discussion; what will constitute evidence of understanding; what are key extensions, connections and generalizations?

This lens helps us make the case for using rich tasks in this thoughtful way, and it engages preservice teachers in some of the thinking that goes into lesson preparation, the focus of our final lens.

Lesson Preparation: Five Questions

Prospective teachers often view lesson planning in a mechanical way that emphasizes format over substantive issues. In our program, rather than speaking of "lesson planning," we use the phrase *lesson preparation* to refer to a generative process that, among other things, produces a *lesson agenda.* In our view, preparing a lesson involves finding answers to these five questions:

i. What are the mathematical concepts about which we want students to construct deep meaning?
ii. What are some tasks that could engage students in deeply understanding these concepts?
iii. What do we have to do to prepare and manage these tasks?
iv. How might these tasks, and our plans to manage them, influence students' mathematical thinking?
v. How will I determine who has learned what about these concepts?

These questions are difficult for prospective teachers to answer, because they have little or no teaching experience or knowledge of students upon which to draw. Lesson analysis using these questions as a lens begins to build that knowledge base.

Analysis of the Pool Border Lesson

The Setup

The Teaching Secondary Mathematics course began with a mathematics "lesson within the lesson" taught by one of us; the prospective teachers in the course were the students. The lesson was built around the "baseball cards" problem:

> Fred laid out his stack of baseball cards in piles of 2, 3 and 4, and had one card left over each time. When he laid them out in piles of 7, there were no cards left over. How many cards did Fred have?

The instructor posed the problem, and the preservice teachers worked in small groups of 3 or 4 until they quickly found that a collection of 49 cards fits all these conditions. A brief whole-class discussion focused on how they found this solution, and then the instructor asked if there were any solutions less than or greater than 49. This led to another period of small-group work followed by another whole-class discussion of the complete set of solutions ($49 + 84n$, where $n = 0, 1, 2, \ldots$), how preservice teachers found this solution set, and the significance of the coefficient 84. Presentations at the chalkboard and ideas offered from their seats drove these discussions, while the instructor asked questions. The instructor then assigned two related problems for homework.

When this 50-minute lesson ended, we asked the preservice teachers to change their roles, from learners participating in the lesson to observers of the lesson, and we used the remainder of this class and the next two classes to introduce the lenses described above and use them, in small groups and as a whole class, to unpack the lessons. First, preservice teachers were given a copy of the instructor's written lesson plan. Using the temporal lens, they identified the "chunks" of the lesson—small-group sessions separated by whole-class discussions—the approximate time each took, and the transitions from one chunk to the next. Second, they examined the dimensions of the lesson, beginning with content analysis of the baseball cards task (including issues raised by the TTLP), and continuing with a look at the learning environment—in particular, the interactions among class participants and between participants and the instructor. Finally, participants were asked to infer from the written lesson plan how the instructor answered for himself each of the five lesson planning questions.

The Pool Border lesson analysis began in week 3. It is built around a case—video excerpts from a 90-minute algebra lesson taught to a class of New York City 8th graders, included in *Learning and Teaching Linear Functions: Videocases for Mathematics Professional Development* (Seago, Mumme, & Branca, 2004). In this videotaped lesson, the 26 children in the class work on a sequence of tasks drawn from the following situation: a square shaped swimming pool is bordered by square tiles, as shown in Figure 1.

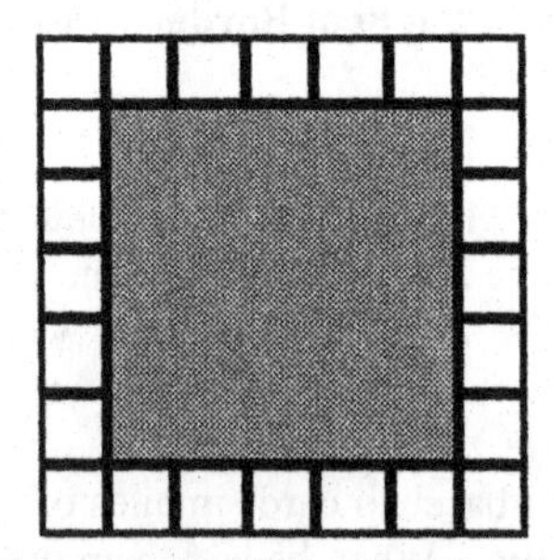

Figure 1. The Pool Border Situation

As described in the next section, the lesson begins with the teacher asking her students, in whole-class format, how many tiles surround this 5-by-5 pool. Once the class agrees that there are 24 tiles around this pool, the teacher asks if there is a way to figure out the number of tiles around a pool of any size. The remainder of this lesson consists of chunks of small-group and whole-class work, in which students develop, offer and try to explain and understand 3 different methods.

We chose these excerpts because they portray a lesson designed to engage children in important mathematical ideas, using structures that are not typical of U.S. middle school classrooms (Hiebert, Gallimore, Garnier et al., 2003). The teacher poses for further investigation questions that are mathematically interesting and important tasks, and she provides opportunities for children to work on these tasks collaboratively. She asks children to explain their thinking, why they think their solutions and strategies are correct, and why the solutions and strategies offered by others are correct. She works hard to make use of children's mathematical thinking throughout the lesson captured on video. However, what makes this a *case*, rather than just a story, is that the excerpts show classroom teaching in all its complexity, and raise many key questions rather than providing authoritative answers (Smith, 2001).

Pre-Viewing Activity

Before we showed preservice teachers the video, we asked them, in a homework assignment, to engage in some mathematical tasks drawn from the pool border situation, and then to outline a 90-minute lesson for an algebra class built around this situation. (See Figure 2.)

1. The questions below refer to the Pool Border situation:
   a. There are 20 tiles around the border of this 4-by-4 pool. How many tiles would there be around the border of an n-by-n pool (where n is a whole number)?
   b. What would be the dimensions of a square pool requiring 288 border tiles?
   c. Explain why there is no square pool with whole number side lengths that would require 190 border tiles.
   d. How many different ways can you think of to count the number of tiles there would be around the border of an n-by-n pool?

2. Outline a 90-minute lesson for a 9th grade Algebra I class, built around the mathematical situation in question 1. You will have to make some assumptions about what your students have been doing and learning in your Algebra I class up to now.

Figure 2. Homework assignment prior to viewing Pool Border video clips

Question 1—familiar to the preservice teachers who explored similar ones in the algebraic thinking portion of the *Mathematics of the Secondary Curriculum* course—was designed to engage our preservice teachers in the mathematics underlying the case. The discussion at the start of the subsequent class generated a wide variety of strategies for counting the number of tiles around an *n*-by-*n* pool, including all of the strategies offered by children in the class on the video.

In their responses to question 2, most of the preservice teachers outlined a lesson similar in structure to that used for the baseball cards lesson we described earlier. (This is not surprising, given the time we spent analyzing that lesson.) During the discussion of these outlines prior to viewing the video, several preservice teachers wondered if such a lesson would fill the 90 minute period.

The Video

Following this homework discussion, we showed several excerpts from the Pool Border lesson. These excerpts comprise about one-third of the lesson, so we provided preservice teachers with what Seago and colleagues (2004) call a "lesson graph," a one-page snapshot of the flow of the lesson, time-coded and showing where the viewer drops in for each of the excerpts. We also provided written transcripts. We watched the excerpts once, and then watched them again with the goal of identifying the mathematical tasks posed to the children in the class throughout the entire lesson. Here are the tasks the preservice teachers identified, and what the children on the tape do with them.

How many unit tiles are there on the border of a 5-by-5 unit square pool? This is a geometry task, and the children generate lots of ideas. One child notes that there are 5 tiles on each side of the pool, and 4 sides, so the number of tiles would be 5 times 4, or 20. The teacher wonders what you are finding when you multiply 5 times 4 in this situation. This generates ideas about area and perimeter. Another child notices that the answer is 24, not 20, and explains that the previous method ignores the 4 corner tiles. Once there is agreement on this answer the teacher poses the next task.

How could you calculate the number of tiles on the border no matter what size pool? With this question, the mathematical focus of the lesson turns to algebra. The class works on this task in groups, and in the subsequent presentation one group offers a generalization of the thinking behind the answer to the first task. They take some time to explain that you can multiply the length of one side of the pool by 4, and then add the 4 corner tiles (See Figure 3.) They summarize with the rule $N = 4S + 4$.

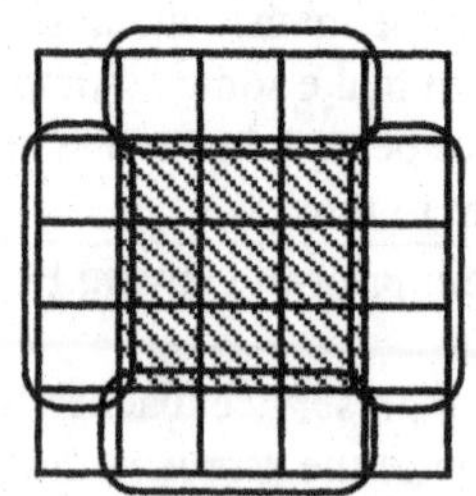

Figure 3. One group's general solution

The teacher builds off this work by asking the following question: *How many ways can you use the picture to come up with ways to count the border tiles?* The children have some difficulty understanding what this task is asking, but after several cycles of small-group work and whole-class discussion, two additional approaches, shown in Figure 4, are proposed.

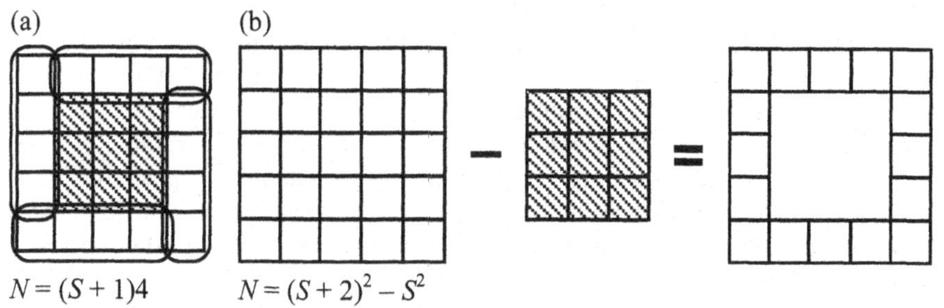

Figure 4. Two additional general solutions

In one solution (Figure 4a), children grouped the tiles along one side of the pool with one of the corner pieces and saw that the border was made up of four of these. In the other solution (Figure 4b), one child found the area of the entire pool with border, and then took away the area of the pool, leaving just the border. Each of these approaches challenged the children to explain their solution in their own words, and challenged the other students to make sense of them. Throughout these whole-class discussions, the teacher posed many questions designed to focus and clarify children's thinking. Then she posed the final task of the lesson: *Are these equations all the same?* As time was running out, one student claimed that all three equations gave the same answer for the 5-by-5 pool that began the lesson. The teacher ended class by asking, "Can you check for equivalency using only one value?"

Post-Viewing Discussions

The four lenses provided the structure for our class discussions and homework assignments following the viewing of the lesson.

Sequence of events. With the lesson graph and a transcript in hand, preservice teachers had some of this work done for them. Thus, our focus was on the structure of each of the chunks, on the length of each chunk, and on how the teacher made the transition from one chunk to the next. For example, the teacher made such a transition when she followed the whole-class discussion of the first solution to the border-of-any-size-pool task with the task of finding other ways to answer this question. Preservice teachers noted how confused many students seemed by what they were being asked to do, and what the teacher did to try to remedy this confusion.

Lesson dimensions. Using this lens, preservice teachers analyzed the student-student and student-teacher interactions they observed. Several times during the lesson, the teacher asked children who were presenting ideas to ask the classmates for questions. This strategy encouraged children to talk to each other. However, most of the time interactions went through the teacher. But the preservice teachers noted that when children did speak to the teacher, she often would respond with questions like "Do you all agree with that?", "Can someone rephrase that or add to that?" and "Does anyone have another idea?" We categorized these questions as re-directions, for the purpose of sending ideas back to the class.

This lens allowed the preservice teachers to focus on the decision to ask students to work in groups and what the teacher had to do to support that work, and the decision to interrupt group work to convene a whole-class discussion. They were able to see and discuss how the teacher used the chalkboard (to record equations offered during presentations, as a place for children to post their work using magnets) and the overhead projector (to pose the initial task and, with the help of colored tiles, to show the solution with corner tiles highlighted in a separate color).

Finally, this lens also allowed us to delve into the concepts of social and *sociomathematical norms* of the classroom (Yackel & Cobb, 1996). Social norms are general patterns of behavior that might be observed in classrooms across subject areas, such as the expectation that you share your thinking with others and listen when others share, and that you explain why you think what you do. Sociomathematical norms, however, are specific to mathematics, for example, not only that you explain your answer, but an understanding in the group of what counts as a convincing mathematical explanation. We focused this discussion with the question, "From the perspective of the children, what does it mean to do mathematics in this class?" This was a difficult distinction for the preservice teachers to make at this point in the semester.

Task analysis (TTLP). As noted in the earlier discussion of the video, the first job for the preservice teachers was to identify the sequence of mathematical tasks posed in this lesson. They were surprised at the number of different tasks that arose out of this single situation, and at the way in which the sequence of tasks moved across the mathematical terrain from geometry to algebra. Then, using the TTLP lens, we dug more deeply into the geometry and algebra underlying these tasks. Specifically,

- How the initial question about the number of tiles around a specific pool raises ideas about perimeter and area for the students on the video;
- That the teacher's question about the number of tiles around any pool asks students to find a function that relates the number of tiles to the dimensions of the pool;
- That this is a linear function, with slope and intercept that can be connected to specific features of the picture (but because these coefficients have the same value, this is more challenging for the students in the video);

- How the students' perceptions of the picture influenced the form of the function they found;
- That the different forms of the function led naturally to the important mathematical question of *equivalence.*

This lens also allowed us to discuss the idea of anticipating student responses to a task, and how this can inform a teacher's thinking about how the lesson might play out. It became clear to the preservice teachers, for example, that the teacher in the case designed a task to elicit a variety of solutions, in part so she could ask if these solutions were equivalent.

The teacher asked many questions during this lesson, and this lens allowed us to characterize them with regard to purpose. Questions like "Do you all agree with that?" and "What do you think of what she said?", which the teacher seemed to use to bring other students into the discussion, were classified by preservice teachers as focusing or clarifying questions. "Can someone else explain what she did?" was viewed as an assessing question, because it helped the teacher to see how many other students understood the mathematical ideas being offered. According to the preservice teachers, the teacher also assessed the understanding of individual students by asking "How did you get that?" several times throughout the lesson. Finally, they saw questions like "How can we take our equation and make it make sense with the picture?" and "Are these equations all the same?" as having the purpose of advancing students' mathematical thinking. Coming into this discussion the preservice teachers knew that asking questions is a good thing, but this framework helped them begin to see why you would want to ask questions in a mathematics class.

Implications for lesson preparation. Coming into this course, the preservice teachers conceived of lesson planning as the process of deciding on a sequence of teacher actions, with little regard for the learners in the room. The five questions that comprise this lens are posed in an order that defers what the teacher will do until after he or she has considered the mathematics and student engagement. We used this lens to ask the preservice teachers to speculate about what answers the teacher might have had to these questions in order to have created this lesson. Using this lens last allowed us to summarize previous discussions about the mathematics, tasks, and instructional strategies and questions, and to model how a teacher might address these questions when preparing a lesson.

While taking part in the set of classes described here, the preservice teachers were also making their first few visits to their field experience classes. The field experience was tied into this case-based lesson analysis through the first of four written field experience assignments, in which we asked them to view and analyze a live lesson using each of the four lenses.

Concluding Comments

Lesson analysis of the kind described here is an important component of our preservice secondary mathematics teacher preparation program. By engaging preservice teachers in the detailed, conceptually coherent study of lessons in which they are learners, lessons they observe (live and on tape), and lessons they help prepare and enact, we bridge the gap between college and school-based parts of our program. By focusing on the lesson as the unit of analysis, we are able to address many aspects of teaching practice, including content and curriculum, learning, teaching, assessment, and context, by bringing one at a time to the foreground for focused study, rather than as separate topics. And using video excerpts as records of practice that can be studied repeatedly gives prospective teachers the opportunity to learn a set of skills and a language that will support their learning throughout their teaching careers.

References

Artzt, A. F. & Armour-Thomas, E. (2001). *Becoming a reflective mathematics teacher: A guide for observations and self-assessment.* Mahwah, NJ: Lawrence Erlbaum Associates.

Cobb, P., & Bowers, J. (1999). Cognitive and situated learning perspectives in theory and practice. *Educational Researcher, 28* (2), 4-15.

Fernandez, C., & Yoshida, M. (2004). *Lesson study: A Japanese approach to improving mathematics teaching and learning.* Mahwah, NJ: Lawrence Erlbaum Associates.

Hiebert, J., Gallimore, R., Garnier, H., Bogard Givvin, K., Hollingsworth, H., Jacobs, J., Chui, A. M. Y., Wearne, D., Smith, M., Kersting, N., Manaster, A., Tseng, E., Etterbeek, W., Manaster, C., Gonzales, P., & Stigler, J. (2003). *Teaching mathematics in seven countries: Results from the TIMSS 1999 video study.* Washington, D.C. National Center for Educational Statistics. Available online: http://nces.ed.gov/timss/Video.asp

Hiebert, J., Morris, A. K., & Glass B. (2003). Learning to learn to teach: an "experiment" model for teaching and teacher preparation in mathematics. *Journal of Mathematics Teacher Education, 6*, 201-222.

Lewis, C., Perry, R., & Hurd, J. (2004). A deeper look at lesson study. *Educational Leadership*, 61 (5), 18-22.

Seago, N., Mumme, J., & Branca, N. (2004). *Learning and teaching linear functions: Videocases for mathematics professional development.* Portsmouth, NH: Heinemann.

Smith, M. S., Bill, V., & Hughes, E. K. (in press). Thinking through a lesson: The key to successfully implementing high level tasks. *Mathematics Teaching in the Middle School.*

Smith, M. S. (2001). *Practice-based professional development for teachers of mathematics.* Reston, VA: National Council of Teachers of Mathematics.

Yackel, E., & Cobb, P. (1996). Sociomathematical norms, argumentation, and autonomy in mathematics. *Journal for Research in Mathematics Education, 27* (4), 458-477.

---

[1]The work reported here is supported, in part, by Teacher Quality Enhancement (TQE), a U.S. Department of Education Title II Partnership Grant (AWARD #P336B040037) awarded to the Metropolitan State College of Denver, Denver Public Schools, and The Fund for Colorado's Future. The opinions, and any errors, are those of the authors alone.

[2]This is based on the results of a survey of AMTE members' methods courses conducted by Robert Ronau, University of Louisville, and P. Mark Taylor, University of Tennessee, and presented at the 2005 annual meeting of AMTE.

[3]Our thinking has been influenced greatly by the work each of us has done in recent years helping classroom teachers engage in adaptations of Japanese Lesson Study (Fernandez & Yoshida, 2004; Lewis, Perry & Hurd, 2004).

Lew Romagnano is Professor of Mathematical Sciences at the Metropolitan State College of Denver, where he leads the Mathematics Teaching and Learning program. He was a Co-Principal Investigator of the NSF-funded research project, Learning to Teach Secondary Mathematics (LTSM), a 5-year longitudinal study of teacher learning. He was also Co-Director of The Interactive Mathematics Program – Rocky Mountain Region (IMP-RMR), and its successor, the Rocky Mountain Mathematics Leadership Collaborative (RMMLC), two NSF-funded projects that provided professional development and leadership support for middle- and high-school teachers and schools changing their mathematics programs. He currently serves on the Mathematical Sciences Academic Advisory Committee of the College Board.

Don Gilmore is an Associate Professor of Mathematical Sciences at the Metropolitan State College of Denver. He works with Teacher Quality Enhancement, a USDOE-funded collaboration with Denver Public Schools, and with the University of Colorado at Denver and Health Sciences Center's NSF funded Rocky Mountain Middle School Mathematics and Science Partnership. The work on these grants involves developing and teaching content courses for preservice and inservice mathematics teachers. His research focuses on professional development for K-12 mathematics teachers with an emphasis on content knowledge for teaching.

Brooke Evans is an Assistant Professor of Mathematical Sciences at the Metropolitan State College of Denver. She teaches and advises students in mathematics and mathematics education, conducts research on mathematics teaching and learning, and engages in a variety of professional service activities. She is the PI for Metro's Mathematics for Rural Schools Program, a web-based professional development project for teachers funded by the Colorado Commission on Higher Education.

Van Zoest, L. R. and Stockero, S. L.
AMTE Monograph 4
Cases in Mathematics Teacher Education: Tools for Developing Knowledge Needed for Teaching
©2008, pp. 117-132

10

Using a Video-Case Curriculum to Develop Preservice Teachers' Knowledge and Skills

Laura R. Van Zoest
Western Michigan University

Shari L. Stockero
Michigan Technological University

We describe our use of a video-case professional development curriculum with preservice teachers in a middle school mathematics methods course. This curriculum was chosen because its video cases build on each other both mathematically and pedagogically, providing opportunities for teachers to make connections among the student thinking in various video clips that are not necessarily possible when using a collection of cases from a variety of sources. We illustrate the nature of the growth in the preservice teachers' mathematical knowledge for teaching and in their development of a reflective stance towards teaching that occurred as a result of their engagement with these materials. We conclude that, with careful adaptations made in response to identified preservice teacher needs, materials developed for use with inservice teachers can effectively be used at the preservice level.

Although the use of cases in preservice teacher education has become increasingly common, the majority of the cases available commercially—both written and video—were developed for use with inservice teachers. We describe our adaptation of one such video-case curriculum—*Learning and Teaching Linear Functions: Video Cases for Mathematics Professional Development, 6-10* [LTLF] (Seago, Mumme, & Branca, 2004)—for use with preservice teachers in a middle school mathematics methods course, and illustrate the nature of the growth in the preservice teachers' mathematical knowledge for teaching and in their development of a reflective stance towards teaching that occurred as a result of their engagement with these materials.

The LTLF Video-Case Curriculum

The LTLF video-case professional development curriculum was developed under a grant from the National Science Foundation to provide professional development for teachers in grades 6-10. The curriculum has a dual focus: (1) to develop teachers' understanding of the mathematics of linear functions; and (2) to develop teachers' understanding of pedagogical issues associated with teaching linear functions. It includes a foundation module, consisting of eight three-hour sessions, and four extension modules, each consisting of two or three three-hour sessions. Extensive facilitation resources accompany each module. As a general sequence of activities, participants individually engage in mathematics tasks as learners and discuss the mathematics as a group before they watch students engaged with the same tasks via one or two short video clips. After a video clip is viewed, participants reflect on the video individually, recording mathematically and pedagogically interesting teaching moments in a journal and using the provided video transcript as a source of evidence to support their observations. They then participate in a whole-group discussion in which participants share the teaching moments they identified and engage with the ideas that others put forth.

Rationale for the Use of LTLF with Preservice Teachers

The middle school mathematics methods course in which the LTLF curriculum is used is the first in a series of three methods courses that Western Michigan University's mathematics department offers for prospective secondary school teachers. This course is required for both mathematics education majors and minors, and is typically taken during students' sophomore or junior year. The LTLF curriculum is used for the first six weeks of the methods course. During the remaining eight weeks of the semester, the preservice teachers participate in group field experiences in local middle school mathematics classrooms, examine and compare middle school mathematics curricular materials, and consider state and national standards and benchmarks for middle school mathematics. For additional details, see the LTLF preservice addenda materials (Mumme & Seago, in development). We incorporated the LTLF curriculum into the course for two reasons: (1) its assumptions and goals are compatible with those of the course; and (2) it provides a coherent set of curricular materials. These reasons are explored in more detail below.

Compatible Assumptions and Goals

The main focus of this NCTM *Standards*-based course is teaching for student understanding by accessing and building on student thinking. We want the preservice teachers to begin to develop the skills, dispositions, and deeper understanding of middle school mathematics needed to teach in this manner. Our assumptions that teaching mathematics is complex and requires a deep understanding of mathematics, and that inquiry and analysis are necessary to improve practice, also guided the development of the LTLF materials. Reflecting these assumptions, the video cases of classroom teaching are intended to stimulate inquiry and reflection, and are not presented as exemplars of teaching. The LTLF curriculum's focus on both mathematics and pedagogy, and the opportunities it provides to focus on student thinking, make it an ideal fit for the course.

Coherent Curriculum

Prior to our adoption of the LTLF curriculum, the cases of teaching used in the methods course were drawn from a variety of sources. Although we kept the focus of each case on student thinking, it was difficult to weave the cases together to build a coherent, integrated curriculum. By design, the videos in the LTLF curriculum build on each other mathematically, providing opportunities to make connections across student thinking in various video clips. An important part of our adoption was the decision to use the entire LTLF foundation module, rather than select a small number of video cases to use periodically in the course. Although this decision resulted in the loss of some content that previously had been part of the course—most noticeably an extended treatment of rational numbers—we felt this loss would be compensated for by what could be gained from engaging preservice teachers in sustained and focused reflection grounded in classroom-based evidence.

Adaptations for the Methods Course

During the methods course, annotated agendas, facilitator notes, PowerPoint® slides, and question prompts included with the LTLF facilitation resources provided the core of the instructional plan. A straightforward adaptation involved reworking the agendas to accommodate the course schedule (two 110-minute class meetings per week). Two more substantial modifications made related to field work and readings are discussed in the following.

Connecting to Practice

The *Linking to Practice* pieces of the LTLF curriculum are designed for teachers to connect what they learn in the sessions to their classroom practice. This is the one area of the curriculum that does not directly translate to work with preservice teachers, although the need to make connections between what teachers learn and their own teaching remains critical. A long-standing component of our middle school methods course is a series of three group field experiences with an intensive reflection component; we modified this experience to capitalize on the preservice teachers' experiences in the LTLF curriculum (see Mumme & Seago, in development, and Van Zoest, 2004, for more details). The field experiences now take place later in the semester, after the preservice teachers have experienced analyzing and discussing student thinking throughout the LTLF foundation module. During these experiences, each preservice teacher facilitates a small group of middle school students as they work to solve a mathematics problem selected directly from the LTLF curriculum. This ensures that the preservice teachers are familiar with the mathematics involved in the problems and have studied different approaches students might take to solve them. These experiences are intended to help preservice teachers further their understanding of how middle school students think mathematically and to provide a valuable opportunity for them to practice their questioning and listening skills.

Required Readings

Although suggested readings are provided in the LTLF facilitation resources, they are not an integral part of the curriculum. We incorporated readings from a range of mathematics education publications, including portions of the NCTM *Standards* (2000), throughout the LTLF foundation module and set the expectation that the preservice teachers would use the readings to inform our class discussions. Requiring these readings seems essential to preservice teacher learning, as they serve as tools to think about learning mathematics in ways other than how the preservice teachers have learned as students themselves, and provide language to discuss such learning. (A subset of the course readings are listed in Figure 1.)

Breyfogle, M., & Herbel-Eisenmann, B. (2004). Focusing on students' mathematical thinking. *Mathematics Teacher*, 97(4), 244-247.

Lannin, J., Barker, D.., & Townsend, B. (2006). Why, why should I justify? *Mathematics Teaching in the Middle School*, *11* (9), 437-443. [This article was not available when the data referred to in the chapter were collected, but has been a part of the course since its publication.]

Phillips, E., & Lappan, G. (1998). Algebra the first gate. In L. Leutzinger (Ed.), *Mathematics in the middle* (pp. 10-19). Reston, VA: National Council of Teachers of Mathematics.

Pirie, S. E. B. (1998). Crossing the gulf between thought and symbol: Language as (slippery) stepping-stones. In Steinbring, H., Bartolini Bussi, M.G., and Sierpinska, A. (Eds.), *Language and communication in the mathematics classroom* (pp. 7-29). Reston, VA: National Council of Teachers of Mathematics.

Scanlon, D. B. (1996). Algebra is cool: Reflections on a changing pedagogy in an urban setting. In D. Schifter (Ed.), *What's happening in math class? Reconstructing professional identities* (Vol. 1, 65-77). New York: Teachers College Press.

Sherin, M., Louis, D., & Mendez, E. P. (2000). Students building on one another's mathematical ideas. *Mathematics Teaching in the Middle School*, *6*(3), 186-190.

Stein, M. K., & Smith, M. S. (1998). Mathematical tasks as a framework for reflection: From research to practice. *Mathematics Teaching in the Middle School*, *3*(4), 268-275.

Wood, T. (1998). Alternative patterns of communication in mathematics classes: Funneling or focusing? In Steinbring, H., Bartolini Bussi, M. G., and Sierpinska, A. (Eds.), *Language and communication in the mathematics classroom* (pp. 167-178). Reston, VA: National Council of Teachers of Mathematics.

Figure 1. Selected readings assigned during LTLF component of methods course

Preservice Teacher Learning

In the remainder of the chapter we turn our attention to what the preservice teachers learned as a result of engaging with the LTLF curriculum. We use classroom transcripts from the spring 2005 semester to focus on two key areas of learning that mirror the dual focus of the curriculum on mathematics and pedagogy.

Learning of Mathematics

We have found the LTLF curriculum to be particularly useful for increasing preservice teachers' *mathematical knowledge for teaching,* described as the mathematics that teachers do in the course of their work (Ball & Bass 2000). We use the delineation of tasks by Ball, Bass, & Hill (2004) as a framework for the kinds of mathematics we want our preservice teachers to learn. Figure 2 describes some ways in which we feel the LTLF curriculum provides opportunities for preservice teachers to develop mathematical knowledge for teaching. In the remainder of this section, we provide a specific illustration of the nature of this learning.

Mathematical Work of Teaching (Ball, Bass, & Hill, 2004, p. 59)	Opportunities to Engage in this Work Provided by the LTLF Curriculum (as Adapted for Preservice Teachers)
• Design mathematically accurate explanations that are comprehensible and useful for students • Use mathematically appropriate and comprehensible definitions • Represent ideas carefully, mapping between a physical or graphical model, the symbolic notation, and the operation or process	These are addressed through the mathematics problems in the LTLF curriculum: • Doing mathematics from the videocases as learners • Discussing their work in the university classroom with high expectations for quality and thoroughness • Focusing on multiple representations and the connections between them
• Interpret and make mathematical and pedagogical judgments about students' questions, solutions, problems, and insights (both predictable and unusual) • Be able to respond productively to students' mathematical questions and curiosities • Make judgments about the mathematical quality of instructional materials and modify as necessary • Be able to pose good mathematical questions and problems that are productive for students' learning • Assess students' mathematics learning and take next steps	These are practiced through engaging with the videocases in the LTLF curriculum and applied in the field experience component: • Watching the videotapes of classroom teaching, using the transcripts to follow-up on hypotheses, and generating hypothetical next moves • Planning a lesson, teaching the lesson, reflecting on what they have learned, and modifying the lesson prior to teaching it to another group of students • Teaching a second lesson to the same group of students • Maintaining high standards of explanation and evidence during class discussions of both the videocases and their own teaching experiences

Figure 2. The mathematical work of teaching afforded by the LTLF curriculum

Our example comes from LTLF Session 6: Regina's Logo. The following dialogue occurred when the preservice teachers discussed their solution strategies after having worked independently on the mathematical problem shown in Figure 3. The written work that was made public during these segments of discussion is provided in Figures 4 and 5.

Assume the pattern continues to grow in the same manner. Find a rule or formula to determine the number of tiles in a figure of any size.

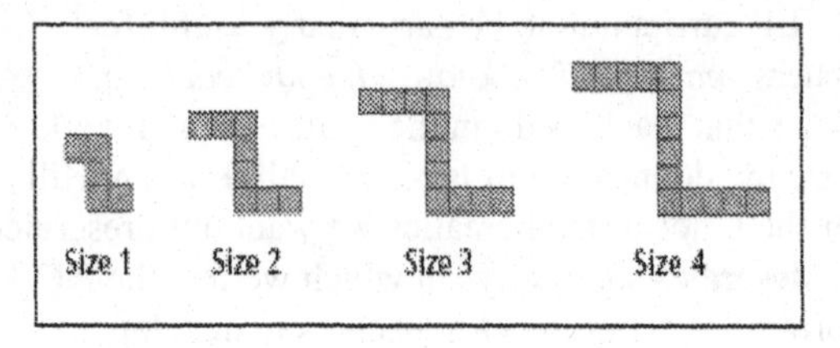

Figure 3. Regina's logo problem
Reprinted with permission from *Learning and Teaching Linear Functions: Video Cases for Mathematics Professional Development* by Nanette Seago, Judith Mumme, and Nicholas Branca. Copyright © 2004 by Nanette Seago, Judith Mumme, and Nicholas Branca. Published by Heinemann, Portsmouth, New Hampshire. All rights reserved.

Laurel

size	# blocks	middle	top (or bottom)	top & bottom
1	5	1	2	4
2	8	2	3	6
3	11	3	4	8
4	14	4	5	10
n		n	$n+1$	$2(n+1)$

$2(n+1)+n$

Steven

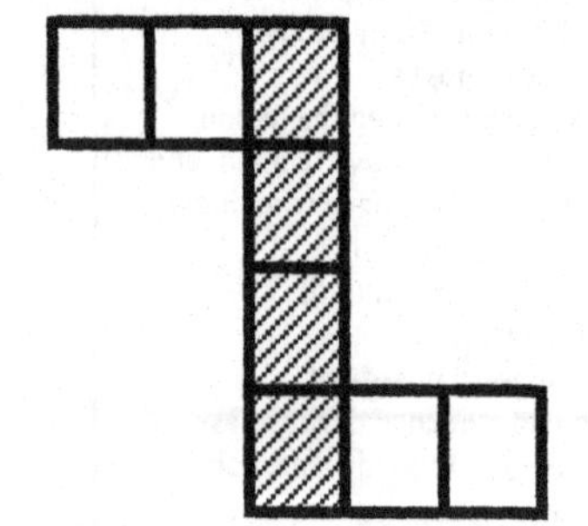

size	middle	top & bottom
1	3	2
2	4	4
3	5	6
4	6	8
	$n+2$	$2n$

Figure 4. Laurel and Steven's work from Regina's logo

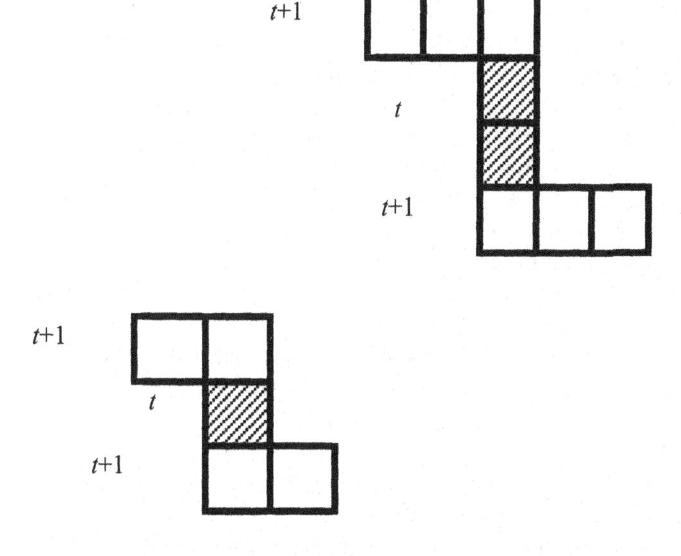

Figure 5. Nathan's work from Regina's logo

In response to a request for someone to share his or her thinking, Laurel[1] presented her approach:

> First thing I did was I made a table to write down the size of each one, it goes to 4. Then, this was the simplest way, I wanted to break it down really simple. Then I just wrote the number of each block in each one. So I just counted it out, it was five, eight, eleven and fourteen. Then, I counted the middle blocks for how many were in the middle for each time. So in the first one there was one, the second one there was two, the third one there was three, the fourth one there was four and I noticed that this and this were the same [pointing to top and bottom rows of squares].

After some clarifications about what she meant in response to questions, she continued:

> Okay. Then I counted the top and bottom, these sides. And in the first one there was two on the top then there was three on the top, then there was four, then there was five, so two, three, four, five. I noticed that this was one plus either one of these 'cause these are the same so I

> noticed that it was $n + 1$ to count for the top and the sides, the top and the bottom, but I know that there's not only a top but there's also a bottom so you want to multiply by 2 and then to account for the middle pieces you're always going to have to add that on as well and that's how I got it, in a nutshell.

At that point the instructor asked, "Anyone have any questions for Laurel?" and Steven and Laurel had the following rapid interchange:

Steven: The only thing I see with that is like what if someone, another student, visualized the middle as like the entire thing [interrupted]

Laurel: Yeah, that's what I wrote in my paper, like that's another way to do it and you'd have to account for someone else [interrupted]

Steven: So basically that would switch the 2 to like $2n + 1 + n + 1$.

Laurel: Yeah, like you could explain it that way, but I see, like what you're saying is that there's more than one way to think of that problem.

At the instructor's request, Steven came to the board and elaborated on his variation of Laurel's solution strategy (see Figure 4). The class then discussed Laurel and Steven's methods and gave them suggestions for clarifying their methods that included this comment from Rick:

> I think it would be better, instead of saying top and bottom for Steven's, wouldn't it be better to say left and right or something like that? Cause even if I didn't have the thinking of Laurel, I would be confused about top and bottom.

After some discussion among the class, the following ensued:

Nathan: [Draws the diagram in Figure 5 on the board] I called this $t + 1$ and this $t + 1$ so I didn't have to say, I didn't have these parts of the chart [pointing to columns], so I didn't have any confusion about the top and bottom. This is $t +1$, this is $t +1$, and then what's left is just t, and then I saw I had two of these, so I had $2(t +1)$ and I had to add my extra t that was in there. And then looking at it the other way [referring to Steven's picture in Figure 4] I started with this and called this $t + 2$ [referring to the shaded column in Steven's picture] and then each of these left over was t so then I had $2t + t + 2$. So that's here and that's that, it's just the same thing, but without the table like that, so I don't have to label my top and bottom or anything.

Instructor: So yours was more visual, just going right directly from the picture.

Nathan: Yeah, it just clears that up, there's no "what do we call the top, what do we call the bottom." Not as much confusion.

Instructor: Anyone have any questions for Nathan? Any comments? [pause] Does that help you think about it better? Is it the same as the table? What do the rest of you think about that?

Mike: I think it's a lot easier to see, personally, the diagram of it, it just seems to make a better explanation for, you know, the middle being just one, you know, the top and bottom being two, you see the information in the chart, but you don't understand why it's one, two, that sort of thing. But with the drawing you can see like the top's $t + 1$, the bottom's $t + 1$, the middle's t, that sort of thing, so you're understanding all the reasoning, like all the numbers.

Although these excerpts convey only small segments of a rather lengthy discussion of the Regina's Logo task, they illustrate the way preservice teachers were able to improve their mathematical explanations and make connections among different representations. Perhaps most importantly, we see a developing awareness of the value of a visual representation and the way in which it can be a tool to understand the mathematics in the situation better. This understanding about visual representations should not be underestimated, as a common misinterpretation of the call for multiple representations in the *Standards* is to have students rotely generate tables, graphs, and symbolic representations with little or no discussion about the connections among them or sense of the benefits offered by different representations.

As is typical, after the preservice teachers had discussed the mathematics in Regina's Logo, they watched the video clip of students discussing this task. While doing so, they hypothesized about what they observed, used the transcript to check the validity of their hypotheses, and then looked for alternate interpretations. The instructor initiated a discussion about their reflections by asking about "mathematically interesting pieces of the video." Thomas responded:

> I think the point where Alexis [a student in the video] talked about using the distributive property to make the equations equal, that one was just a simplified version of the other was important to show that she was recognizing that each equation went at it a different way but it was really the same thing, that they were both getting to the same result and answer, even though they were looking at it a different way.

The discussion that followed included this interaction:

Instructor: Why was it important that she recognize that they were equal in this case?

Rick: Because there's two different methods of looking at the problem. You know, both of them can be represented with an equation, both equations can be manipulated to be visually equal, visually as well as algebraically equal.

Instructor: [after a pause] And we're supposed to be doing the building [trails off; referring to a classroom norm of building on each other's comments, as described in Sherin, Louis, & Mendez (2000)]

Reed: In taking that one step farther, she possibly recognizes that it's going to be easier to formulate an equation using the closed method instead of recursively, and then that will help with the next part of the lesson where it's, Schemel's Logo [the next problem in the LTLF curriculum, which the preservice teachers had completed as homework], which it looks like it would be pretty hard to figure out recursively.

Here we see that, with prompting, the preservice teachers are developing an understanding of what it means to acknowledge students' mathematical thinking, make sense of it, and determine whether it is likely to be productive. Furthermore, they are engaging with mathematical ideas (e.g., distributive property, recursion, equality) that are fundamental to their future students' learning.

One thing that has become clear to us through watching our students engage in these opportunities to develop mathematical knowledge for teaching is the difficulty of both developing and implementing such knowledge. One advantage of the LTLF curriculum is that it is a coherent curriculum and is designed to develop understanding over time. Through the field experiences, we have seen that implementing this knowledge while teaching requires another level of skill and expertise—one that is best developed through continued interaction with students. Because of this ongoing development of mathematical knowledge for teaching, it is critical that preservice teachers develop a reflective stance for teaching—one that will enable them to learn from their ongoing teaching experiences.

Development of a Reflective Stance

We have documented three important shifts in how preservice teachers who engaged with the LTLF curriculum analyzed instances of practice: (1) a shift from evaluating instructional decisions based on affective measures to analyzing instruction based on pedagogical considerations; (2) a shift from knowing about student thinking to conjecturing about student thinking; and (3) a shift towards considering how instructional decisions might affect student thinking (Stockero, 2006). Each of these shifts is illustrated and briefly discussed.

Shift from affective to pedagogical considerations. Early in the semester, the preservice teachers tended to consider instructional decisions based on affective measures, as is seen in the following excerpt from LTLF Session 1: Growing Dots, in which Mandi discusses Kirk's (a teacher in the LTLF video) decision to push a student (James) to go to the board to explain his solution:

> I don't think that the teacher should have like made him go up there. Especially at that time, like kids are just in adolescence and they're like, it's a rough time and they don't want to be put on the spot in front of the whole class. So like, he didn't want to go up there and the teacher was like, big deal, come up here anyway. Like, I don't think he should have made him come up if he didn't want to.

In this passage, Mandi is disagreeing with Kirk's decision based on how it might have made James feel, rather than considering pedagogical reasons that Kirk may have had for encouraging James to share his thinking. For example, James had solved the problem recursively and had arrived at a different solution than a student who had shared her explicit expression for solving the same problem—raising important mathematical issues that Kirk may have wanted other students to consider.

Throughout the semester, the preservice teachers increasingly began to consider pedagogical and mathematical reasons for instructional decisions. For example, in a discussion of LTLF Session 6: Regina's Logo, Cathy comments on a teacher's (Gisele) less central role in her classroom. In the video clip, Gisele is standing off to the side of the classroom, while two students are at the board recording other students' solutions and asking questions of their classmates.

> I just think she takes it to, like, too much of an extreme. If I was a kid who didn't understand, I would tune out…I don't know which one knows what they're talking about, so I'll just tune out until we get to the end and maybe somebody will explain what's going on.

Although Cathy's response is a reaction based on how she might feel as a student, it also shows some consideration for student learning. More significantly, other preservice teachers responded to Cathy's comment by offering opposing opinions that were based on pedagogical considerations, such as the comment by Thomas:

> If you need your teacher to constantly tell you at what point the mathematical idea is important, at what time do you move from the teacher being the one of authority to maybe one of your peers, one of the students coming up with an idea that's important and recognizing, on your own, that that's a good idea, an important mathematical concept? I don't know if the teacher is always saying "well this is important, this is important," that you would begin to make your own

growth and be able to identify, by yourself, the important mathematical concepts, which I think is important as you get later in education and more into the real world.

Counter-responses, like that made by Thomas, did not occur early in the semester. Thus, this shift from focusing on affective measures to pedagogical considerations represents an important change that took place during the class discussions.

Shift in interpreting student thinking. Early in the semester, the analyses of student thinking tended to be absolute in nature—not fully considering the reasoning behind a students' inaccurate or incomplete response, yet using definitive language. Later in the semester, however, the preservice teachers became more tentative in their analyses of student thinking.

Consider, for example, the following dialogue from LTLF Session 1, Growing Dots, focused on James' thinking. James was considering the number of dots in a pattern that began with a single dot in the center and added four dots around it each second.

Rick: Didn't James get 400 instead of 401. So he just added wrong then, basically?
Mike: Yeah.
...
Rick: So, that's what he did though, right? He just simply added wrong? He just looked for, I mean, he probably didn't have enough time to add those...
Theresa: to 400, he probably just saw that he didn't have the one like she did and said like, "oh, well I got 400". I mean I think it would have taken him a long time to add to...
Rick: Right, exactly, that's the point.
Theresa: I mean he didn't do it all the way out, he just said, "Oh, well I'm wrong because I got 400."

In this segment, Rick assumes that James came up with an incorrect answer of 401 because of an adding error and both Mike and Theresa support his analysis. In the video, however, James clearly explained why he got 400 dots instead of 401 like his classmate: "'Cause she counted the center, we didn't count the center like she did." This type of analysis is characteristic of comments made early in the semester—instead of turning to evidence from the transcript, these preservice teachers made ungrounded assumptions about James' thinking that were not countered by others in the class.

Discussions of student thinking in the LTLF videos became richer and more grounded later in the semester, particularly after class discussions centered on a portion of the *Standards*, and on articles related to classroom discourse. The excerpt below from LTLF Session 6, Regina's Logo, illustrates this change. This dialogue builds on conjectures that had been shared about a student's (Jordan) understanding:

Laurel: Ok, so like maybe he just got confused, what, like, his wording, his vocabulary at times meant...[the teacher] should have just like kind of slowed him down and then like, "Think about your vocabulary. You know, are you using the right words to describe what you're trying to get across?"...

Mandi: That relates [to] the discourse articles that we read, like how the whole mathematical language and, like there's the one example that was $2x$ and she kept saying x squared; it was really two x and the teacher was saying to them you have to say $2x$, not x squared....

Mike: I don't think it has to do with vocabulary, I think it has to do with him just misunderstanding how to do it because like he talks about, you know, at first he says three x is x times x times x and then later on he also says that x plus two is equal to two x.

There is an important difference between this discussion and the discussion of James' thinking earlier in the semester. Rather than immediately conclude that Jordan didn't understand the mathematics in the problem, the preservice teachers considered an alternative explanation—language—for the errors in Jordan's thinking. In addition, they challenged each other's ideas, rather than accept the first explanation offered. This suggests that the preservice teachers were moving away from a definitive stance of *knowing* about student thinking to a more tentative stance in which they *made conjectures* based on the evidence available. As Ball and Cohen (1999) have argued, this tentative and inquisitive stance is a central quality of teachers who continually learn as they teach, and thus are able to adapt their teaching in response to students' understandings.

Considering how instructional decisions affect student thinking. In early class discussions, the preservice teachers tended to talk about what they "liked" or "didn't like" about the instructional decisions that a teacher made, rather than discuss how those decisions affected student learning. This changed, however, later in the semester. Consider, for example, an excerpt from a class discussion following a field experience in which Hillary considers how her classmate's instructional decision might have hindered student thinking. In this case, the teacher, Vince, had introduced the term *variable* early in his session with the students.

> I think also on doing it that way, Evan might have been a little bit thrown off by the teacher saying "let's put a variable in there" if he wasn't ready for that yet. Cause on line thirteen, Vince says, "Let's just throw in the variable t. What's t?" You know and Evan is still trying to explain, but he wasn't thinking variables, he was thinking well I'm just

> doing top, bottom, and sides. I don't know what you mean, you know...So that could have been maybe a little bit of Evan's frustration at the beginning cause he wasn't sure how to express it in variables. He just knew how, what he was seeing.

Rather than simply state that she didn't like the teacher's actions, Hillary reflects on how an instructional decision affected student thinking, providing an example of the increased number of preservice teachers' reflections focused on pedagogy and student thinking. This, along with the dialogue presented in the preceding sections, provides evidence of important shifts toward an increased focus on student thinking that took place using the LTLF curriculum (see Stockero, 2006, for more detailed analyses). These shifts are significant, as teachers who become more aware of student ideas through the use of videos subsequently pay more attention to such ideas in their own teaching practice (Borko & Putnam, 1996; Sherin, 2000, 2004).

Conclusions

The preservice teacher learning reported here suggests that curriculum materials developed for use with practicing teachers can be effectively used at the preservice level. In particular, the LTLF professional development materials provided a coherent curriculum that allowed us to strengthen the course's main focus—teaching for student understanding by accessing and building on student thinking. Prior to using the LTLF video curriculum, the middle school methods course had used a collection of cases drawn from a variety of sources. Although we focused class discussions on the student thinking in each case, it was difficult to connect the cases together so that they built on each other to form a coherent curriculum. The LTLF cases solved this dilemma because they were intentionally designed to connect and build on each other both mathematically and pedagogically.

It is important to remember, however, that the learning goals for the course and the curriculum materials were compatible, and the curriculum was used intact. Making substantial adaptations to the curriculum, such as using only some of the video segments, would require careful planning and consideration of the consequences of such adaptations. The preservice teachers with whom we worked did not begin to show significant changes in their reflection until a month into the LTLF curriculum (Stockero, 2006), suggesting that sustained engagement was critical.

We conjecture that the adaptations we did make—incorporating field experiences and requiring readings—were essential to support the preservice teacher learning we have seen. With limited experiences on which to draw, providing alternative perspectives via course readings was an important scaffold in the preservice teachers' learning and needs to be considered when adapting materials designed for inservice teachers for use at the preservice level. Furthermore, structuring the field experiences so that they built on the video cases—mathematically and in terms of the reflection that was required—also seemed to support the preservice teachers' first attempts at reflecting on their own practice to recognize their strengths and their areas for continued work.

The challenging nature of teaching makes it essential that teachers enter the field with a foundation of mathematical knowledge for teaching and a reflective stance that allows them to learn from their ongoing teaching experience. Our work provides support for the use of a video-case curriculum as a means to develop a reflective stance in preservice teachers and further their understanding of the mathematics needed for teaching.

References

Ball, D. L., & Bass, H. (2000). Interweaving content and pedagogy in teaching and learning to teach: Knowing and using mathematics. In J. Boaler (Ed.), *Multiple perspectives on mathematics teaching and learning* (pp. 83-104). Westport, CT: Ablex Publishing.

Ball, D. L., Bass, H., & Hill, H. C. (2004). Knowing and using mathematical knowledge in teaching: Learning what matters. In A. Buffler & R. C. Laugksch (Eds.), *Proceedings of the 12th Annual Conference of the Southern African Association for Research in Mathematics, Science and Technology Education*. Durban: SAARMSTE.

Ball, D. L., & Cohen, D. K. (1999). Developing practices, developing practitioners. In L. Darling-Hammond & G. Sykes (Eds.), *Teaching as the learning profession: Handbook of policy and practice* (pp. 3-32). San Francisco: Jossey-Bass.

Borko, H., & Putnam, R. T. (1996). Learning to teach. In D. C. Berliner & R. C. Calfee (Eds.), *Handbook of educational psychology* (pp. 673-708). New York: Simon and Schuster Macmillan.

Mumme, J., & Seago, N. (in development). *Developing facilitators of practice-based professional development*, NSF Project #0243558, WestEd.

National Council of Teachers of Mathematics. (2000). *Principles and standards for school mathematics*. Reston, VA: Author.

Seago, N., Mumme, J., & Branca, N. (2004). *Learning and teaching linear functions: Video cases for mathematics professional development, 6-10*. Portsmouth, NH: Heinemann.

Sherin, M. G. (2000). Viewing teaching on videotape. *Educational Leadership, 57*(8), 36-38.

Sherin, M. G. (2004). New perspectives on the role of video in teacher education. In J. Brophy (Ed.), *Using video in teacher education* (pp. 1-28). Oxford, UK: Elsevier Ltd.

Sherin, M. G., Louis, D., & Mendez, E. P. (2000). Students building on one another's mathematical ideas. *Mathematics Teaching in the Middle School, 6*(3), 186-190.

Stockero, S. L. (2006). *The effect of using a video-case curriculum to promote preservice teachers' development of a reflective stance towards mathematics teaching.* (Doctoral dissertation, Western Michigan University, 2006). *Dissertation Abstracts International, 67* (09A), 3339-3520.

Van Zoest, L. R. (2004). Preparing for the future: An early field experience that focuses on students' thinking. In T. Watanabe and D. R. Thompson (Eds.), AMTE Monograph 1: *The work of mathematics teacher educators* (pp. 124-140). San Diego, CA: Association of Mathematics Teacher Educators.

---

[1]All names are pseudonyms expect for those in the LTLF videos.

Laura R. Van Zoest, Professor of Mathematics Education at Western Michigan University, received her Ph.D. from Illinois State University. Through university courses, school-based professional development, and initiatives funded by the National Science Foundation and Michigan Department of Education grants, she has supported inservice and preservice teachers as they implement the recommendations of the NCTM *Standards.* Her research questions revolve around how one becomes an effective teacher of mathematics and the role of learning communities in the process. She is editor of the NCTM book, *Teachers Engaged in Research: Inquiry into Classroom Practice, Grades 9-12.*

Shari L. Stockero, Assistant Professor of Mathematics Education at Michigan Technological University, completed her Ph.D. in mathematics education at Western Michigan University in 2006. She previously taught mathematics and physics at the high school level. Her dissertation research focused on the use of video cases of teaching in preservice teacher education, specifically the extent to which such materials can be used to develop a reflective stance towards teaching. Her current research focus is on how to support mathematics teacher learning effectively at both the preservice and inservice levels.

Goldsmith, L. T. and Seago, N.
AMTE Monograph 4
Cases in Mathematics Teacher Education: Tools for Developing Knowledge Needed for Teaching
©2008, pp. 133-145

11

Using Video Cases to Unpack the Mathematics in Students' Thinking[1]

Lynn T. Goldsmith
Education Development Center, Inc.

Nanette Seago
WestEd

This chapter reports an investigation of ways that a video-case based curriculum, in concert with skilled facilitation, can promote shifts in the ways teachers use video cases to consider issues of students' algebraic understanding. Eight teachers participated in a 12-session professional development seminar for middle- and high-school teachers that used the Learning to Teach Linear Functions: Video Cases for Mathematics Professional Development *materials. The first and eighth sessions of the seminar involve analysis of the same video clip. Analysis of the group discussions from these two sessions indicate that the later viewing yielded a more extensive, complex, and elaborated discussion than the one that took place five months earlier. Teachers' mathematical analyses of the students' thinking were more detailed and robust, and focused on students' potential for further learning rather than a narrow evaluation of what the students did or did not know. Although the written seminar materials were designed and structured to provide opportunities for teachers to develop deeper understanding of students' algebraic thinking, we believe that the facilitator also played a central role in helping teachers to focus on the key ideas emphasized in the curriculum.*

Although teacher educators have been enthusiastic about using cases (and other forms of artifact-rich professional development) as a teaching tool for inservice and preservice teachers, researchers are just beginning to investigate learning that artifact-rich professional development can promote

(Kazemi & Franke, 2003; Masingila & Doerr, 2002; Nikula, Goldsmith, Blasi, & Seago, 2006; Seago & Goldsmith, 2006; van Es & Sherin, in press). The lack of knowledge has been largely a function of timing: before the research community could begin to investigate teachers' learning in case-based professional development settings, the field had to create the phenomenon. With a variety of commercial products now available (for example, Barnett, Goldstein, & Jackson, 1994; Schifter et al., 1999a, 1999b; Seago, Mumme, & Branca, 2004; Smith et al., 2005), researchers can begin to investigate teachers' learning. Two important dimensions are emerging in this work: developing mathematics knowledge for teaching and focusing teachers' attention.

(1) *Developing mathematics knowledge for teaching.* Researchers are beginning to report that artifact-based professional development helps develop the mathematical knowledge necessary for teaching. By pairing exploration of teachers' own mathematical solutions to problems with study of cases of students' work with the same problems, teachers strengthen the mathematical understanding needed to interpret student work (Borko, 2006; Kazemi & Franke, 2003; Seago & Goldsmith, 2006). In part, this involves learning to go beyond an assessment of whether a student's work is correct to following (or speculating about) the mathematical ideas behind student's work. For example, how does the student's solution provide clues about how the student seems to understand certain mathematical ideas? Are there mathematical inconsistencies in the work? How do the representations the student uses provide clues to his/her understanding? As teachers consider these questions in their work with classroom artifacts, they can develop a more detailed and articulated ("unpacked," or "decompressed") understanding of the mathematical ideas under consideration, and can also make connections to related ideas (Ball, Hill, & Bass, 2005; Ferrini-Mundy, Floden, McCrory, Burrill, & Sandow, 2004).

(2) *Focusing teachers' attention.* A second dimension of teachers' learning in artifact-rich professional development involves learning to attend to critical features of mathematics teaching and students' mathematical thinking and learning. Teachers must suspend their tendency to evaluate the surface features of the work captured in the artifact (e.g., Angel's work is incorrect, Betsy spent a lot of effort color-coding her solution, the teacher shouldn't have asked James to come to the board if James said he didn't want to). Instead, teachers must consider the student and teacher thinking *behind* the activity the artifact captures. In part, this shift of attention requires teachers to learn to attend to different aspects of the classroom (and the curriculum) than they may have traditionally considered (e.g., what Angel *did* understand, how the color coding might have helped Betsy represent and solve the problem, what the teacher knew about James's work and why he might have wanted James to come to the board). The notion that professional expertise involves the development of particular ways of organizing and attending to the environment runs through a variety of different disciplines, from archeology to chess to teaching

(Goodwin, 1994; deGroot, 1965; Mason, 2002; Nussbaum, 1990; Sherin, 2001).

What might teacher learning within these two dimensions look like? As part of the work of the Turning to the Evidence (TTE) project, we have been studying how a video-case based curriculum, in concert with skilled facilitation, can promote shifts in the ways teachers use video cases to consider issues of students' algebraic understanding (Nikula et al., 2006; Seago & Goldsmith, 2006). In the remainder of this chapter we briefly describe the professional development context we studied and illustrate some of the changes we observed in teachers' analysis of one video case at two points in time.

The Professional Development Context

The professional development seminar took place in the Attwood Unified School District during the 2002-2003 academic year. Attwood is an urban district on the west coast and serves approximately 18,000 students. The student population is primarily Hispanic and Caucasian. Student test scores were in need of improvement, and district administrators were eager to find professional development opportunities for mathematics teachers in the three middle- and two-high schools. Nine teachers participated in the seminar.[2]

The seminar used *Learning to Teach Linear Functions (LTLF): Video Cases for Mathematics Professional Development* (Seago, Mumme, & Branca, 2004) as the professional development curriculum. To ensure that the materials were implemented with fidelity, Nanette Seago, lead LTLF author, facilitated the group.

LTLF is a modular curriculum intended for use with teachers in grades 5-10. The foundation module is eight sessions long and focuses on unpacking ideas of slope and y-intercept, exploring issues of indexing and rate of change, and emphasizing the interconnectedness of different representational forms. Each session builds on previous work, focusing on a set of interconnected algebraic concepts and representational forms relating to linear functions. The first and last sessions both use the video case "Growing Dots 1." Revisiting the case at the end of the module provides an opportunity for participants to reconsider the case in light of the work they have done together. Subsequent modules focus on deepening and extending ideas developed in the foundation module and include modules launching a lesson or task, interpreting and responding to unexpected student methods, making use of student ideas during discussion, and examining equivalence and generalization.

Each seminar session begins by situating the day's work in relation to the previous session and discussing teachers' thoughts and observations about a "linking to practice" assignment between seminars. This is followed by the group's exploration and discussion of the mathematics problem used by the teacher in the upcoming video case. This mathematical work includes solving the problem for themselves and forecasting alternative solutions, comparing different solutions, and anticipating students' misconceptions. The mathematical work is intended to prepare teachers for a focused examination of the mathematics as it unfolds within the classroom video clip.

Unpacking the Mathematics Behind Students' Thinking: Two Discussions of James and the Growing Dots

In both Sessions 1 and 8, teachers viewed and discussed a video segment of 9th grade students presenting solutions to the Growing Dots problem (see Figure 1). Prior to viewing the video in Session 1, the teachers worked on the problem themselves, sharing and discussing their own solution strategies.

At the beginning At 1 minute At 2 minutes

Describe the pattern. Assuming the sequence continues in the same way, how many dots will there be at 3 minutes? 100 minutes? *t* minutes?

Figure 1. Growing Dots Problem

The video segment captures the beginning of the whole group discussion of the problem in which two students, Danielle and James, share their solutions for the number of dots at $t = 100$ minutes. Danielle uses a closed solution to describe the growth. "I got the equation *x times 4 plus 1*. The *plus 1* being the center. . *x* being the dots around it, or 100, and *4* being all the dots except the center." At the request of the teacher (Kirk), she illustrates her solution at the board. When she is done Kirk asks, "Can I get somebody who maybe sees this a little bit differently?" James then describes a recursive solution.

> I wasn't worried about this [dot] in the center. I didn't really think of that as 1. . . . [The problem] says like 'the sequence' or whatever. Well, this [first stage] started at 1 and over here [the second stage] is 5, and 9 over here. I just took. . .they added four every time. See like four, four, four. And that's why I got *x plus 4* for the equation."

In both Sessions 1 and 8 of the LTLF seminar, Nanette began the discussion of the clip by asking participants to identify "mathematically interesting and important moments."

Session 1. This conversation opened with a comment about James' inattention to the proper use of variables, and quickly moved from discussion of James' work to a general observation about "what kids do." Bruce observed,

> I just think that [an issue is] identifying the variable again. I just think it's one of the common mistakes that the students don't read what the question says. When they get to looking at a problem, sometimes the directions go out the window and they're just looking to see what it's doing and not answering the question.

Walter expressed puzzlement at James' solution, and several teachers interpreted James' description of x ("the number that I was adding it to in the previous picture") as simply an idiosyncratic oversight on his part. The teachers had difficulty connecting James' approach with their own ways of thinking about the closed solution, $4x + 1$.

Janice: So his variable is not consistent throughout. The variable changes from the beginning to minute 1 to minute 2. It's not a constant variable. It changes. . . .

Walter: It's kind of curious. How did he get the 400 if he didn't know the one before? Cause he needed . . . to know the one before to add 4.

No one responded to Walter's question. Instead, there was a brief discussion about James' decision to ignore the (unchanging) center dot, in which Bruce observed that Kirk had missed an opportunity to work on vocabulary: "it seems to me [this] would be a good . . .place for him [Kirk] to bring up this term *constant.*" Walter, however, still seemed puzzled by James' work and again raised the issue of how James got his answer.

Walter: (consulting transcript) At 19:10 [transcript time code] James tries to explain where he got x plus 4. . . . Kirk is trying to get him to explain what is x and he never really did.

Tom: Cause that would have been 104. . . .

Annie: But he got 400. I can't figure out how he got 400 with his formula. . . .

Janice: The only thing I can think of, yeah, is that he, he drew the picture.

Tom: I think he had the picture in his mind. He could see that it was growing and so

Annie: But his formula didn't match his answer.

Tom: Right. And we're obviously going to get that in every class. Kids are going to see it differently.

Annie: And that's like what we were saying about how a kid will pick up a formula and he'll never check it.

The teachers were engaged in the conversation, identifying James' focus on the growing part of the pattern as mathematically interesting. They did not dismiss his work as wrong (in fact, they specifically noted later in the conversation that he wasn't actually incorrect), but the teachers were not sure of his reasoning and did not pursue the question of how his notation and his numerical answer were connected.

Throughout the discussion, Nanette's facilitation kept the discussion focused by asking participants to clarify their comments, requesting that they ground comments in evidence from the video transcripts, paraphrasing and checking to ensure they followed one another's observations, and encouraging them to listen to and respond to each other. She also paraphrased participants' comments to highlight the mathematics in the idea and redirected the group's attention. For example, the group focused entirely on James's solution until Nanette asked directly about Danielle. Their response to her query was to evaluate rather than explore Danielle's solution, which they saw as being like their own. Although they noted that her notation would have been confusing without hearing her explanation, they were not particularly inclined to examine it for insights into the details of her thinking or as a contrast or foil for James' ideas.

Nanette: What about Danielle? Did anybody. . . ?

Janice: I thought she did a great job with that.

Tom: She had, she has an A on it.

Annie: At first I didn't know what she was talking about, about 4, and she drew a little "4" on the end? . . . Like, 1, 2, 3, 4. That was the only thing I didn't understand. . . .I mean, I knew what she was trying to do, I just didn't. . . .

Janice: Right. If I had walked in off the street and looked at what she had drawn, not having been there for her explanation, I would never have been able to figure out how she got her answer. Because I would have thought that would have been 4 times *four*. {Laura: four, four, and four, yeah.} Not 1 times 4.

The teachers began their work together in this seminar with a tendency to judge the *correctness* of the work captured in the video, rather than to examine the *nature* of students' thinking.

Session 8. When the group revisited the video five months later, teachers were generally more engaged in analyzing James' response and far less puzzled by his solution. Absent were comments like those from Session 1, which focused on inattention to the assignment as the source of students' errors or offered suggestions for promoting vocabulary learning. Instead, the group worked to make sense of James' thinking in its own right. This second conversation was more mathematically coherent as well, in that the ideas the teachers raised were tightly interconnected and built on each other more than in the earlier discussion. In addition, the teachers spent time trying to unpack James' strategy, focusing on what he understood rather than on what was problematic about his solution.

Janice: Go back to the lesson graph. It says, "Kirk posts the task on the board. Describe the pattern." . . . James wasn't all that wrong. He just didn't look at it more long term. . . .

Laura: He is describing it. . . Look at the beginning, you know, as x, plus 4—. . . that's a pattern. You look at minute 1, that's x, plus 4, and he got to minute two. So he's not all that wrong. Now, he, he's not thinking broadly enough, but. . .

Trevor: He'd gotten so deep into the table that he forgot the part that starts the table going.. . . if you just looked at lines 2, 3, and 4, you'll see a change taking place for sure, but that doesn't give you your initial conditions because you're seeing the changes taking place as you march down the table. . . .

Walter: He thought that if he counted the center one, you would be counting something that had not been added; and his interpretation was, 'how many had been added? What's the pattern?' . . . What's the pattern. Yeah—he answered, 'what is the pattern?'. . .

Sally: He was looking at what was generated off the center.

Trevor: He was looking at the growth.

The discussion also reflected the group's willingness to consider James' solution from James' own mathematical perspective, rather than assume he was either careless or fell short of being able to produce the expected, closed notational form. And in trying to understand James' ideas, the group was able to see the potential for building on his ideas to help him develop a deeper understanding of the linear relationships represented in the problem.

Sally: [James was] just seeing pieces of it. . . .

Walter: Yeah. {several others: uh huh} He was looking here, said 'plus 4' . . . .

Janice: He was looking at each step as an independent. . . as something independent of the previous.

Trevor: He was looking at the step process and not seeing the entirety. . . .

Janice: As soon as James got from *point A* to *point B*, he forgot all about *point A*. And now *point A* became *point B* . . . and he moved on. . . .

Nanette's facilitation of this conversation was also different from her facilitation of Session 1; in the intervening six sessions, Nanette helped to focus the group on analyzing both their own mathematical thinking and the mathematical ideas that seemed to be at play among the students in the video cases. By Session 8, Nanette was able to facilitate a more substantive conversation on mathematical issues. The first session had been a time for her to get a feel for the group and the ways they thought about the mathematics and James and Danielle's approaches to the problem. In the intervening months, the teachers explored a number of representations of linearity (including a series of variants of the Growing Dots problem) in a series of structured activities that focused on articulating the connection between geometric and symbolic representations of the slope and *y*-intercept. These activities included teachers' own mathematical explorations as well as viewing classroom video that emphasized work with different representations. As facilitator, Nanette's work involved highlighting important ideas that participants raised and posing questions and ideas that they may not have quite been ready to pose on their own.

As the seminar work progressed and participants began to dig deeper into the mathematics and mathematical thinking behind linear functions, Nanette continued to focus participants on more exacting mathematical analyses, highlighting and expanding on their spontaneous analyses of students' mathematical thinking and focusing their attention on specific mathematical moments in the video clips. By Session 8, the teachers had become more analytical about the elements of a linear function and were prepared to see the internal logic of a seemingly incorrect solution.

The discussion from Session 8 continued with further consideration of James' thinking, in response to a question Nanette posed about how James was thinking about the variable in his expression.

Nanette: What was his x?

Janice: His x was the number of dots added at each minute. Or—no, no; x was the previous picture.

Walter: So that's why he was looking at just one thing.

Nanette: So the one thing he was looking at was the *plus 4*.

Walter: The previous to the next one. Plus 4.

Laura: That's exactly what it was. So . . .every time he's adding four.

. . .

Walter: x plus 4. x plus 4. x plus 4. I think that's a great way to lead into the recursive method. . . 'All right James, how much is it now?' 'Plus 4.' 'OK, next one?' 'Plus 4.' And there's your formula.

Nanette: So, so, could you expand on that? What [do] you mean by that? . . .

Walter: I'd just kind of walk through with James how many times we'd add a four. . . . So, kind of help him step backwards a little bit and see that it's not just talking about what happened from the second to the third step, or the fourth to the fifth step, but to see this thing as Danielle did from further back . . .

Nanette: What would you help him see? How many fours are being added?

Walter: Different amount. Start with something small first, and then keep on adding more of them so he can kind of step back slowly. . . .So one step backwards at a time, let him take a couple of more steps, say 'hey there's a whole bunch of these. . . .' 'OK, this one was plus 4, this was plus 4, and this was plus 4. How many did we add all together from here to here?'

Somewhat later in the discussion, Nanette connected Walter's discussion of the "plus 4" recursion to Trevor's earlier comment about the initial condition, highlighting the different aspects of linear growth they had examined over the course of the seminar. A little later in the conversation, she refocused Walter's observation about helping James see the center dot as part of the pattern of growth, posing it as an issue of understanding the initial state, which is the constant.

Nanette: You're talking about then helping him see *how* many fours are being added, cause therein lies the help in seeing that there are fours added each time. And then I thought what Trevor was raising was a little bit different, which was also helping him see the initial state. So at "time zero" there's one dot. . . looking at that as the beginning. So, one *plus* how many fours? That's different than $4x$ plus 1—that's one plus x fours. Is that what you're saying? . . .So you're, so representing what James is doing or helping James see something a little bit closer to a generalized form, but going from where he's thinking. . . .

Walter: [looking at the growth] would help him understand that the center *is* part of the pattern.

Nanette: Or potentially related to what Trevor was saying, 'what does it start with?' What are you starting with?

Trevor: It's the stone that goes in there, causes the ripple.

Nanette made connections here between Walter's discussion of the "plus 4" recursion and Trevor's earlier observation that James ignored the initial condition. Thus, she highlighted different components of linear growth that the group was examining and also reinforced James' mathematical potential—that James has ideas about the pattern of growth, which a teacher could use to help him develop a generalized means of representing the change.

At the end of the discussion Walter explored this idea, and Nanette used this opportunity to highlight subtle but potentially interesting differences between ways students might represent the growth. Walter noted,

> Instead of extending [the pattern], you could retract it and say, 'how far can you retract it?' And [James] would probably say, 'Oh, that's the very first one you could have.' And then you could kind of help him see that 1 was indeed part of it.

Nanette responded, taking Walter's observation one step further.

> Potentially, it might be order. Instead of adding the 1 at the end as the rest of the class appeared to be working on, it could be starting with the 1. Which seems like it's the same to all of us, but it could be different to kids. Not adding it back in; *starting* with it might be different.

Trevor, still thinking about initial conditions, added, "So, *b* plus *mx*." Nanette wanted the group to continue to consider order, and answered, "*b* plus ***xm***"! Finally, Walter related these two similar, but subtly different representations, returning the conversation to issues of practice and "repacking" the mathematics by noting, "And then you can come talk about why those both are really equivalent."

Conclusion: No Case is an Island

The Session 8 conversation about Growing Dots was more extensive, complex, and elaborated than the one five months earlier. There is more detail to teachers' observations about James' treatment of the pattern of growth and more speculation about his reasoning for not counting the center dot. Teachers talked about what both Danielle and James "saw" in the problem, and recognized the potential for James to gain even deeper understanding. With Nanette's guidance, teachers compared and contrasted different ways of representing the growth, and considered the psychological differences among these different representations. All in all, the teachers seem to have learned to unpack the mathematics underlying James' thinking and to tie his explanation to fundamental mathematical ideas about linear relationships.

We are currently in the process of tracing the work of this group of teachers over the intervening months in order to draw a clearer picture of how the mathematical stances of the participants (and of the group as a whole) developed over time. Although much of the impetus for their development is embodied in the LTLF materials themselves, we must emphasize that their *implementation* is central to the outcome (Nikula et al., 2006; Seago, 2007; Stein, Smith, Henningsen, & Silver, 2000). In many ways, Nanette was a *more* active member of the group in Session 8 than in Session 1. She pressed more for participants to clarify their understanding of James' mathematical ideas and connected threads of the conversation to help teachers make mathematical connections. Even the most carefully structured curriculum materials cannot anticipate the particularities of the ideas that will arise in professional development groups, and an important job for the facilitator is to keep activities and conversations focused on the goals of the professional development. As a field, we have yet to focus on the facilitation demands of professional development. One place to begin investigating the work required to facilitate the use of cases (or any classroom artifact) effectively is to consider the kinds of moves that characterize facilitation (Remillard & Geist, 2002; Sassi & Nelson, 1999) and the strategies that facilitators use to focus teachers' attention on analyzing mathematically important aspects of video cases (Nikula et al., 2006). For example, we believe that facilitators can help teachers learn to ground their analysis of mathematics learning and teaching in the evidence from the video itself. Over the course of their work together, Nanette would ask participants to identify places in the video or transcript that supported their observations and analyses. Sometimes she made a request to encourage teachers to speak about specific instances of mathematical thinking they noticed, rather than to talk about students' thinking in general terms; other times she asked for evidence when she thought there were alternative plausible interpretations and wanted to open the issue up for further discussion. Other tasks of facilitation might include helping teachers learn to develop a stance of curiosity about the student thinking captured in artifacts rather than taking an evaluative, "grading" approach; looking for strengths in students' thinking, even in cases where a solution wasn't totally successful; using a guiding mathematical framework to analyze the artifact; and developing a plausible story line to account for all of the thinking captured in an artifact.

References

Ball, D., Hill, H., & Bass, H. (2005). Knowing mathematics for teaching: Who knows mathematics well enough to teach third grade, and how can we decide? *American Educator*, February, 14-17; 20-23; 43-46.

Barnett, C., Goldenstein, D., & Jackson, B. (Eds.). (1994). *Fractions, decimals, ratios, and percents: Hard to teach and hard to learn?* Portsmouth, NH: Heinemann.

Borko, H. (2006, April). Learning from the problem solving cycle. Paper presented at the annual meeting of the American Educational Research Association. San Francisco, CA.

Ferrini-Mundy, J., Floden, R., McCrory, R., Burrill, G., & Sandow, D. (2004). A conceptual framework for knowledge for teaching school algebra. Unpublished manuscript. East Lansing, MI: Michigan State University.

Goodwin, C. (1994). Professional vision. *American Anthropologist, 96*, 606–633.

deGroot, A. (1965). *Thought and choice in chess.* The Hague: Mouton.

Kazemi, E., & Franke, M. (2003) *Using student work to support professional development in elementary mathematics.* Center for Study of Teaching and Policy: University of Washington.

Masingila, J. O., & Doerr, H. M. (2002). Understanding pre-service teachers' emerging practices through case studies of practice. *Journal of Mathematics Teacher Education, 5* (3), 235-263.

Mason, J. (2002). *Researching your own practice: The discipline of noticing.* London: Routledge Falmer.

Nikula, J., Goldsmith, L. T., Blasi, Z. V., & Seago, N. (2006). A framework for the strategic use of classroom artifacts in mathematics professional development. *NCSM Journal of Mathematics Education Leadership, 9* (1), 57-64.

Nussbaum, M. C. (1990). The discernment of perception: An Aristotelian conception of private and public rationality. *Love's knowledge: Essays on philosophy and literature.* Oxford: Oxford University Press.

Remillard, J. T., & Geist, P. K. (2002). Supporting teachers' professional learning by navigating openings in the curriculum. *Journal of Mathematics Teacher Education, 5*, 7-34.

Sassi, A. M., & Nelson, B. N. (1999, April). Learning to see anew: How facilitator moves can reframe attention. Paper presented at the meetings of the American Educational Research Association, Montreal, Ont.

Schifter, D., Bastable, V., Russell, S. J., Lester, J. B., Davenport, L. R., Yaffee, L., & Cohen, S. (1999a). *Building a system of tens: Facilitator guide.* Parsippany, NJ: Dale Seymour.

Schifter, D., Bastable, V., Russell, S. J., Lester, J. B., Davenport, L. R., Yaffee, L., & Cohen, S. (1999b). *Making meaning for operations: Facilitator guide.* Parsippany, NJ: Dale Seymour.

Seago, N. (2007, Winter). Fidelity and adaptation of professional development materials: Can they co-exist? *NCSM Journal of Mathematics Education Leadership, 9* (2), 16-25.

Seago, N., & Goldsmith, L. T. (2006). Learning mathematics for teaching. In J. Novotna, H. Moraova, M. Kratka, & M. Stehlikova (Eds.), *Proceedings of the 30th conference of the International Group for the Psychology of Mathematics Education.* Vol 5, pp. 73-80, Prague: PME.

Seago, N., Mumme, J., & Branca, N. (2004). *Learning and teaching linear functions.* Portsmouth, NH: Heinemann.

Sherin, M. G. (2001). Developing a professional vision of classroom events. In T. Wood, B. S. Nelson, & J. Warfield (Eds.), *Beyond classical pedagogy: Teaching elementary school mathematics* (pp. 75-93). Hillsdale, NJ: Erlbaum.

Smith, M. S., Silver, E. A., Stein, M. K., Henningsen, M. A., Boston, M., & Hughes, E. K. (2005). *Improving instruction in algebra: Using cases to transform mathematics teaching and learning, Volume 2.* New York: Teachers College Press.

Stein, M. K., Smith, M. S., Henningsen, M. A., & Silver, E. A. (2000). *Implementing standards-based mathematics instruction: A casebook for professional development.* New York: Teachers College Press.

van Es, E. A., & Sherin, M. G. (in press). Mathematics teachers "learning to notice" in the context of a video club. *Teaching and Teacher Education.*

---

[1]This paper reports on the work of the Turning to the Evidence project, which is funded by the National Science Foundation as grant REC-0231892. The findings and opinions expressed in this article are not necessarily those of the National Science Foundation. Project members include Mark Driscoll, Johannah Nikula, and Zuzka Blasi. The authors would like to thank them for their participation in all aspects of the work. In addition, we would like to thank Barbara Nelson and Ilene Kantrov for their contributions to the ideas in this paper.

[2]Eight were middle school teachers (five women and three men) and one (Trevor) taught 9th grade. One of the middle-school teachers (Bruce) had to drop the course due to conflicts; the remaining eight completed the 12 seminar sessions.

Lynn Goldsmith's interest in mathematics education grew from research focusing on the development of mathematically talented students. Her work has since expanded to considerations of central issues in educational reform from a variety of perspectives: students' learning, teachers' instructional approaches, and the support curriculum materials provide for both teaching and learning. Co-PI of the Turning to the Evidence project, she is co-author of *Nature's Gambit: Child Prodigies and the Development of Human Potential* (Teachers College Press), senior author of *Choosing a Standards-based Mathematics Curriculum* (Heinemann), author and series co-editor of the *Guiding Middle-grades Curriculum Decisions* series (Heinemann), and director of research for the recently published K–5 mathematics curriculum, *Think Math!* (Harcourt School Publishers).

Nanette Seago currently serves as Co-PI for three NSF projects, the Turning to the Evidence research project and two materials development projects, *Developing Facilitators of Practice-Based Professional Development* and *Learning and Teaching Geometry: Video Cases for Mathematics Professional Development*. Since 1998, Seago has served as the Project Director and Co-PI for the Video Cases for Mathematics Professional Development project (VCMPD), funded by the National Science Foundation to develop professional development curriculum materials. In 2002, Seago collaborated with LessonLab in the development of the TIMSS-R public release videos and the design of an online course sponsored by Intel Corporation entitled *TIMSS Video Studies: Explorations of Algebra Teaching*. She is lead author of *Learning and Teaching Linear Functions: Video Cases for Mathematics Professional Development, 6-10* (Heinemann).

[illegible]

Appendix
AMTE Monograph 4
Cases in Mathematics Teacher Education: Tools for Developing Knowledge Needed for Teaching
©2008, pp. 147-148

Appendix

Bibliography of Mathematics Cases

Boaler, J., & Humphreys, C. (2005). *Connecting mathematical ideas: Middle school video cases to support teaching and learning.* Portsmouth, NH: Heinemann.

Barnett-Clarke, C., & Ramirez, A. (Eds.). (2003). *Number sense and operations in the primary grades: Hard to teach and hard to learn?* Portsmouth, NH: Heinemann.

Barnett, C., Goldenstein, D., & Jackson, B. (Eds.). (1994). *Fractions, decimals, ratios, and percents: Hard to teach and hard to learn?* Portsmouth, NH: Heinemann. (Two books: case book and facilitator's guide)

Bastable, V., Schifter, D., & Russell, S. J. (2002). *Examining features of shape.* Parsippany, NJ: Dale Seymour Publications. (Two books: case book and facilitator's guide, video)

Bush, W. S. (Ed.). (2000). *Mathematics assessment: Cases and discussion questions for grades 6-12.* Reston, VA: National Council of Teachers of Mathematics.

Bush, W. S., & Dworkin, L. (Eds.). (2001). *Mathematics assessment: Cases and discussion questions for grades K-5.* Reston, VA: National Council of Teachers of Mathematics.

Merseth, K. K. (2003). *Windows on teaching math: Cases of middle and secondary classrooms.* New York: Teachers College Press. (Two books: case book and facilitator's guide)

Miller, B., Moon, J., & Elko, S. (2000). *Teacher leadership in mathematics and science.* Portsmouth, NH: Heinemann.

Russell, S. J., Schifter, D., & Bastable, V. (2002). *Working with data.* Parsippany, NJ: Dale Seymour Publications. (Two books: case book and facilitator's guide, video)

Schifter, D., Bastable, V., & Russell, S. J. (2008). *Reasoning algebraically about operations.* Parsippany, NJ: Dale Seymour Publications. (Two books: case book and facilitator's guide, video)

Schifter, D., Bastable, V., & Russell, S. J. (2007). *Patterns, functions, and change.* Parsippany, NJ: Dale Seymour Publications. (Two books: case book and facilitator's guide, video)

Schifter, D., Bastable, V., & Russell, S. J. (2002). *Measuring space in one, two, and three dimensions.* Parsippany, NJ: Dale Seymour Publications. (Two books: case book and facilitator's guide, video)

Schifter, D., Bastable, V., & Russell, S. J. (1999). *Making meaning of operations*. Parsippany, NJ: Dale Seymour Publications. (Two books: case book and facilitator's guide, video)

Schifter, D., Bastable, V., & Russell, S. J. (1999). *Building a system of tens*. Parsippany, NJ: Dale Seymour Publications. (Two books: case book and facilitator's guide, video)

Seago, N., Mumme, J., & Branca, N. (2004). *Learning and teaching linear functions: Video cases for mathematics professional development, 6 – 10*. Portsmouth, NH: Heinemann.

Smith, M. S., Silver, E. A., Stein, M. K., Boston, M., Henningsen, M. A., & Hillen, A. (2005a). *Improving instruction in rational numbers and proportionality: Using cases to transform mathematics teaching and learning, Volume 1*. New York: Teachers College Press.

Smith, M. S., Silver, E. A., Stein, M. K., Henningsen, M. A., Boston, M., & Hughes, E. K. (2005b). *Improving instruction in algebra: Using cases to transform mathematics teaching and learning, Volume 2*. New York: Teachers College Press.

Smith, M. S., Silver, E. A., Stein, M. K., Boston, M., & Henningsen, M. A. (2005c). *Improving instruction in geometry and measurement: Using cases to transform mathematics teaching and learning, Volume 3*. New York: Teachers College Press.

Stein, M. K., Smith, M. S., Henningsen, M. A., & Silver, E. A. (2000). *Implementing standards-based mathematics instruction: A casebook for professional development*. New York: Teachers College Press.

Wilcox, S. K. & Lanier, P. E. (Eds.). (2000). *Using assessment to reshape mathematics teaching*. Mahwah, NJ: Lawrence Erlbaum. (Print materials and video)

Printed in the United States
By Bookmasters